Ecosee

Ecosee

Image, Rhetoric, Nature

Edited by
SIDNEY I. DOBRIN
and
SEAN MOREY

Published by
State University of New York Press, Albany

Printed in the United States of America

For information, contact State University of New York Press, Albany, NY
www.sunypress.edu

Production by Diane Ganeles
Marketing by Fran Keneston

Library of Congress Cataloging-in-Publication Data

Ecosee : image, rhetoric, nature / edited by Sidney I. Dobrin and Sean Morey.
    p. cm.
  Includes bibliographical references and index.
  ISBN 978-1-4384-2583-2 (hardcover : alk. paper)
  ISBN 978-1-4384-2584-9 (pbk. : alk. paper)
  1. Environmental policy. 2. Ecology. 3. Imagery. 4. Rhetoric.
5. Nature study. I. Dobrin, Sidney I., 1967– II. Morey, Sean, 1979–

HC79.E5E289 2009
333.7—dc22
                                                        2008020819

10 9 8 7 6 5 4 3 2 1

This one's for Carla Blount, because for eleven years she has been there for me, making sure I get my work done, and I can never say "thank-you" enough for that.

—S. I. D.

To my parents, Frank and Susan, who never told me what to think but instead let me see for myself.

—S. M.

Contents

Figures

Acknowledgments

I offer gratitude and thanks to Priscilla Ross, James Peltz, Larin McLaughlin, and the entire staff at State University of New York Press for their continued support of my work. Much thanks to Sean Morey for the collaboration on this project and the discussions and ideas the collaboration fed. I am grateful for the friendship and collegiality of Raúl Sánchez, Joe Hardin, and Julie Drew, who push me to think about my work; without them, none of this would ever get done. Sincere thanks to M. Jimmie Killingsworth and Jacqueline S. Palmer for their willingness to contribute the Afterword to this collection and for their continued support and friendship. Sean and I also offer our gratitude to Jackie Palmer's Texas A&M University Tech Writing 320 class, which provided incredible copyediting feedback on early drafts of *Ecosee*; we are especially grateful to student editors Kara Hinesley, Pam Soderstrom, Megan Speck, Holly Hudgins, Blake Coleman, Nancy Elliott, Jennifer Kalinowski, Holly Southerland, Sarah Graham, Lauren Davis, Amy Noesser, and Natalie Bell. I also am grateful for the opportunities provided to me by the Department of English at the University of Florida and for my colleagues' support. I am forever indebted to John Leavey, Pamela Gilbert, Kenneth Kidd, Jack Perlette, Phil Wegner, Susan Hegeman, Kim Emery, Marsha Bryant, Greg Ulmer, Robert Ray, Ed White, Malini Schueller, Terry Harpold, Ira Clark, Bob Thomson, Judy Page, and Jim Paxson for their friendship and for providing an atmosphere that encourages dialogue and learning, even in trying times. Their work inspires me to do more with mine. Thanks also to Carla Blount, my program assistant, who, as I said in the Dedication, makes sure I get my work done. My deepest thanks and love to Teresa, Asher, and Shaia, who are my world and without whom none of what I do would matter. Finally, thanks to the contributors to this collection for their time and energy in putting this book together; Sean and I are both grateful to you all for your efforts.

—S. I. D.

I am grateful to Andrew M. Gordon for providing comments on earlier drafts of the Introduction and chapter 1, "A Rhetorical Look at Ecosee." I thank Greg Ulmer, Raúl Sánchez, Clay Arnold, Lindsey Collins, and Carol Steen for providing a network through which to conduct some of the ideas here. Thanks to my family (Frank, Susan, Tim, Andy, Julie, Jess, Nick, Megan, Sarah, Emma, and Kelsey) for providing another kind of network, to Cathy Bester, Rob Robins, Coach Wise, who keep my mind off writing, and Jesse Brown (my daimon, who always stays silent). To the nonhuman animals that I know well: Dusty, Milo, Skye, and Toss. And, finally, my appreciation to the Zaffke family and to Aubrey, who sees that the writing gets done and keeps me going.

—S. M.

Ecosee: A First Glimpse

Sidney I. Dobrin and Sean Morey

My first view—a panorama of brilliant deep blue ocean, shot with shades of green and gray and white—was of atolls and clouds. Close to the window I could see that this Pacific scene in motion was rimmed by the great curved limb of the Earth. It had a thin halo of blue held close, and beyond, black space. I held my breath, but something was missing—I felt strangely unfulfilled. Here was a tremendous visual spectacle, but viewed in silence. There was no grand musical accompaniment; no triumphant, inspired sonata or symphony. Each one of us must write the music of this sphere for ourselves.

—Charles Walker, U.S. astronaut

In the Introduction to their 1992 book *Ecospeak: Rhetoric and Environmental Politics in America*, M. Jimmie Killingsworth and Jacqueline S. Plamer initiate a conversation about ecospeak, "a makeshift discourse for defining novel positions in public debate" about environmental and ecological issues (8). Killingsworth and Palmer explain that "like Newspeak, the austere vocabulary of mind control in Orwell's politicolinguistic fable *1984*, ecospeak becomes a form of language and a way of framing arguments that stops thinking and promoting cooperation through communication" (9). That is, ecospeak operates to establish political capital without calling into question its own position, its own politics. For Killingsworth and Palmer, ecospeak is a rhetorical object in need of critical examination in order to break the hold of ecospeak by identifying various discourses on the environment before they are galvanized by dichotomous political rhetoric. *Ecospeak: Rhetoric and Environmental Politics in America* does so by studying the transformations of these discourses as

1

they enter the public realm by a local discourse community (whether a professional ghetto like "the scientific" community or an actual region defined by geographic and democratic features). At the very least, such an analysis can reveal possible identifications and real conflicts passed over by an ever-too-glib retreat into ecospeak (10).

Ecospeak, in detail, then, performs a rhetorical analysis of a number of works by writers "representing several distinct ethical and epistemological perspectives on environmental issues" (11). The book itself is first an act of rhetorical analysis of a particular kind of discourse that Killingsworth and Palmer have aptly termed *ecospeak*. However, the sophistication of their study contributes to a larger conversation about how ecology, environment, and even nature are formed by and through discourse in which ecospeak can be seen not (only) as a particular discursive object in need of the analysis performed by Killingsworth and Palmer but instead as the larger framework that identifies that rhetoric—in the case of *Ecospeak*, writing in particular—and the politics of environment and ecology are inextricably bound.

Ecospeak is a work of both discourse analysis and discourse theory, and as such it is a monumental work. However, *Ecospeak* does not take into account the role images play in promoting various ecospeaks, nor in the larger examination of the relationships between discourse and environment/ecology. Of course, it would be unfair to claim that this is a failure of *Ecospeak*, as the new media boom of the late twentieth and early twenty-first centuries had only begun to take hold when *Ecospeak* was published. *Ecosee: Image, Rhetoric, Nature* works beyond Killingsworth and Palmer's attempt to understand "the relationships among language, thought, and action in environmental politics" to take into consideration the visual facet of environmental rhetoric. Ecosee, then, is the study and the production of the visual (re)presentation of space, environment, ecology, and nature in photographs, paintings, television, film, video games, computer media, and other forms of image-based media. Ecosee considers the role of visual rhetoric, picture theory, semiotics, and other image-based studies in understanding the construction and contestation of space, place, nature, environment, and ecology. Ecosee is not (only) an analysis of existing images, it is a work toward making theories that put forward ways of thinking about the relationship between image and environment, nature, and ecology, as well as a theory (or, more accurately, a number of theories) of visual design for those who make images. Ecosee is bound to writing, as the production and interpretation of image walk hand in hand with the production and interpretation of written discourse. While Killingsworth and Palmer rightly identify that "as much as the environmental dilemma is a problem of ethics and epistemology, it is also a problem of discourse" (6), so too is the environmental dilemma(s) a problem of image/imaging. For ecosee, though, the environmental dilemma is not just a political/ecological crisis about the protection of the environment but a dilemma of representation, a dilemma of rhetorical and visual-rhetorical choice.

The environmental movement, which has taken various forms since its modern inception in the late 1960s, has sparked wide scholarship on the ways that messages about the environment are communicated. Such approaches toward this study usually include environmental rhetoric, environmental discourse, or, more recently, ecocomposition. These subject areas, usually housed in departments of history, political science, the natural sciences, communications, and English, focus on the language used by both environmentalists and anti-environmentalists[1] and how this language becomes coded and appropriated by all sides of eco-political struggles. However, these studies traditionally have paid little attention to how images are used to spread eco-political capital and how these "eco" images might interact with texts and other images. While scholars have successfully focused on the verbal/discursive representations of nature and the environment, they have, for the most part, overlooked its visual representation and construction.

Of course, other disciplines, such as art history, have certainly developed traditions of research that address visual representations of nature. It would, for instance, be impossible to address the works of painters such as George Catlin, Thomas Moran, Albert Bierstadt, and Winslow Homer without some attention to their representations of landscape. The same could be said of Georgia O'Keefe, whose paintings represent not only landscape but shells, rocks, bones, and flowers. To attempt to list artists—whether painters or otherwise, known or unknown—who have created works that represent nature would be impossible. Likewise, a number of works have taken up the examination of the relationship between art, image, and representations of nature. For instance, E. H. Gombrich's classic *Art and Illusion: A Study in the Psychology of Pictorial Representation* examines relationships between the imitation of nature and the role of tradition. Gombrich points out early in his masterpiece that "artists know that they learn by looking intensely at nature, but obviously looking alone has never sufficed to teach an artist his trade" (1960, 11). Gombrich goes on to develop a theory of mimesis throughout *Art and Illusion* that deeply examines the traditional relationship between art and nature. Though he identifies nature as an ideology, he poses an argument for the "naturalness" of imagery. Likewise, theorist and critic W. J. T. Mitchell's landmark books *Picture Theory: Essays on Verbal and Visual Representation* (1994) and *Iconology: Image, Text,*

Ideology (1986) invoke Gombrich's work to establish some of the most groundbreaking theories of images of late. In developing such theories, Mitchell addresses the representation of nature in a number of ways: the relationship between nature and illusion (*Picture Theory*), the role of the self in seeing nature (*Picture Theory*), nature versus convention (*Iconology*), and nature as imitation (*Iconology*). Similarly, John Berger's books *About Looking* (1980) and *Ways of Seeing* (1977) work to understand the relationships between image and reality, arguing that "all images are man-made" and that "when we 'see' a landscape, we situate ourselves in it" (1977, 9, 11). Previous work by Steve Baker, a contributor to this book, has also set the tone for a contemporary evaluation of the relationship between art/image and nature. Baker's *The Postmodern Animal* (2000) is one of the most captivating studies of how contemporary art (exemplified in the work of Olly and Suzi—who he takes up in his contribution to this collection as well—Mark Dion, Damien Hirst, Sue Coe, and a number of others) works not only to represent nature but to shape the very idea of identity. In his earlier work, *Picturing the Beast* (2001), Baker examines the role of animal images in contemporary culture, developing a theory of "disnification" in which the image and representation of animals are often reductive, presenting animals as stupid, trivial, and of limited value. With these works and others in mind, we can easily identify that various disciplines of artistic production also have developed a scholarly history of examining and producing representations of nature. Yet few have done so with an extended agenda of examining the politics and (visual) rhetorics of those images (Berger's and Baker's works are notable exceptions, ones that are critical to the foundations from which *Ecosee* evolves). Within and beyond this tradition, *Ecosee* works to bring together a range of disciplinary works to coalesce various efforts to better understand the role of image and visual representations of nature in constructing the politics of nature and environment.

The study of nature's visual representation is particularly important given that a large part of individuals' experiencing nature involves seeing nature as nature.[2] Much of the rhetoric evoked by environmentalists or nature enthusiasts is that of the visual expanse of nature: grand vistas, crystal-clear waters, resplendent flora and fauna. One recognizes this as well in the writings of John Ruskin:

> This first day of May, 1869, I am writing where my work was begun thirty-five years ago, within sight of the snows of the higher Alps. In that half of the permitted life of man, I have seen strange evil brought upon every scene that I best loved, or tried to make beloved by others. The light which once flushed

those pale summits with its rose at dawn, and purple at sunset, is now umbered and faint; the air which once inlaid the clefts of all their golden crags with azure is now defiled with languid coils of smoke, belched from worse than volcanic fires; their very glacier waves are ebbing, and their snows fading, as if Hell had breathed on them; the waters that once sank their feet into crystalline are now dimmed and foul, from deep to deep, shore to shore. These are no careless words—they are accurately—horribly—true. I know what the Swiss lakes were; no pool of Alpine fountain at its source was clearer. This morning, on the Lake of Geneva, at half a mile from the beach, I could scarcely see my oar-blade a fathom deep. (1903–1912, Preface, Vol. 19, pg. 293)

Ruskin describes the declining quality of his environment due to pollution from nearby factories, but this passage is not interesting only because of what it says about threats from pollution but what it says about how the environmentally concerned understand nature in two ways. First, Ruskin shows us how those deeply concerned for the environment feel a pressing need to write about it. They need to actively and discursively construct their idea of nature, and here Ruskin compares two states of his environment at two different times. However, he also shows us how we discursively construct not just a general picture of nature but the picture itself. Ruskin employs visual cues such as various colors, sunsets, mountains, light, and crystalline. He also uses visual verbs such as sight, seen, see. Perhaps because of his work as an art critic, Ruskin knows what his lakes were through the visual, and he knows that the environment is healthy when it is clear and clean.

However, we must make this clear: although Ruskin uses imagery to create a textual picture of his environment, he does not present an image. As Gorgias argues in Plato's dialogue of the same name: "To begin with, he does not say a color, but a saying" (1989, 980 b 5), identifying a fundamental difference between the words articulated as a description of the color and the color itself. Plato's point, and one that Jean-François Lyotard echoes, is that we can never know the object in the world but can only address it and understand it through language. Despite the term *imagery*—as it is used in poetry—the imagery of language is not a visual image. It may rely upon the metaphor of sight and convey images within the mind, but ten words in a poem will necessarily omit the other 990 signifiers that real images can convey.

Lyotard also explains that we construct reality through language: "Reality is not what is 'given' to this or that 'subject,' it is a state of the

referent (that about which one speaks) which results from the effectua-
tion of establishment procedures defined by a unanimously agreed-upon
protocol" (1988, 4). The reality of "nature" is similarly an agreed-upon
social construction that humans often take for granted as "real." There is
no "nature" that exists in the world except as a discursively constructed
concept. Again quoting Lyotard: "Even in physics, there exists no protocol for establishing the reality of the universe, because the universe is
the object of an idea" (5). Just as "reality is not a given" (9), nature
is not a given but must be established through language, whether that
language includes the verbal, visual, or both.

Although perhaps true of most of our daily interactions, our interaction with nature is inherently visual; most of our outdoor activities
rely on sight for their engagement. One visits the Grand Canyon to
experience its visual vastness, and one hikes along the Florida Scenic Trail
for its scenery. Signs to such parks and recreation areas often enforce
this visual interaction: "Leave only footprints, take only photographs."
Activities such as photography, fishing, or hunting all require the visual
for their participation and enjoyment, and even the tools used to carry
out these activities reflect this: a camera lens, a fishing lure that seeks
to visually mimic natural prey, attached to the line by the hook's "eye,"
a hunting rifle's "sight."

It is not surprising then that *Homo sapiens*' first artwork and
writing depicted nature. Many scholars have pointed to the caves at
Chauvet-Pont-d'Arc as evidence of the human propensity toward the
representation of nature. Many see these depictions as first examples of
art, and others identify them as precursors to writing. In either case,
their importance grows from the relationship between the need for
visual representation—either art or writing—and the need to represent
nature. With this first art/writing comes a human visual construction of
nature. What is the rhetorical significance of the fact that compared to
the art/writings found in other regions of France, the caves at Chauvet-
Pont-d'Arc depicted dangerous animals, while "the animals most often
depicted in Paleolithic caverns are the same as those that were hunted"
("Time and Space")? Could this suggest that even within the same
area different people valued, through representation, different parts of
nature over others? Did one group represent nature because of its use
as food, and another because of its potential danger? The representation
of nature itself becomes a rhetorical representation, one that constructs
a reality of nature. A difference in representation suggests a difference
in ideological construction. Composing nature through images does not
represent that nature, but composes, making the image an ontological
surface below which the real was never present.

Given the historical importance of images in constructing nature, it is little wonder that environmental groups have incorporated images into their rhetorical strategies, and that they rally around constructed icons. Robert Gottlieb points out that "more than many social movements, environmentalism has become associated with compelling ideas and images—whether Nature (the value of wilderness) or Society (the negative associations of urban pollution or hazards)" (2001, 5). These images do not become passively associated with any particular environmental idea or political movement but are actively incorporated into the agenda of such groups because of the images' rhetorical qualities, based in the pathos, ethos, or logos of an image, or a combination of the three. However, some images are so shocking that they almost instantly become iconic on their own. Maarten Hajer explains this about the representation of planet Earth:

> If there is one image that has dominated environmental politics over the last twenty-five years it is the photo of the planet Earth from outer space. This picture, which entered the public imagination as an offspring of the 1960s Apollo space programme, is said to have caused a fundamental shift in thinking about the relationship between man and nature. The confrontation with the planet as a colorful ball, partly disguised by flimsy clouds, and floating seemingly aimless in a sea of utter darkness, conveyed a general sense of fragility that made people aware of human dependence on nature. It facilitated an understanding of the intricate interrelatedness of the ecological processes on planet Earth. Indeed, the image, it is said, caused a cognitive elucidation through which the everyday experience of life in an industrialized world was given a different meaning. (1995, 8)

Like Ruskin, Hajer shows us the Earth (or in this case, a representation of the Earth) through language rather than including the photograph in his book. He describes Earth's colors, shape, and features to provide his reader with a verbal picture of the planet. So although Hajer claims that the image was so powerful, and he points this out at the beginning of his work, the written word gains preference over the image. Of course, unlike Paleolithic Neanderthals, we no longer rely solely on images as material media to convey meaning but have transitioned to written text (writing, of course, being a form of image, though we skew that distinction here to indicate an artificial difference between writing and other forms of image); Ruskin's and Hajer's depictions make clear that if we want to understand how pictures represent the environment, then we must come to them through a textual explanation. As Gottlieb also explains, "these images are made manifest by language and representation" (2001, 5).

If these images are so important to both Hajer and Gottlieb, then why do both abandon their discussion of the image after just one reference? Both authors quickly turn from image and its impact on the perception of the environment to language instead. If images are so powerful in how we construct environment, as Hajer points out with the image of the Apollo Earth, then scholars should focus beyond environmental rhetoric and discourse as primarily language based and also look at it as image based. They should examine how the environment creates images, and how these images create the idea of the environment. This is the project of ecosee: to study the visual representation of nature and environments in photographs, paintings, television, movies, video games, and all forms of new media that use images. Such a study theorizes how humans use images to construct ideas of nature and environment, how those images reinforce those constructions, and how humans may use existing images (or make new ones) to create alternative ways of seeing nature and environment. Theories of ecosee consider how and what images—both the idea of the image and specific images themselves—might suggest about the environment and also look toward a variety of perspectives from different disciplines—visual semiotics, environmental rhetoric, image theory, spatial theory, ecology, to name a few—and their elements that theories of ecosee might contain.

One can almost hear the grumbling now: first ecocriticism, then ecospeak, next ecocomposition, and now ecosee. Two neopests (Gregory L. Ulmer's neologism for those who needlessly create neologisms) are at it again, making up another empty word upon which to build a book. And this we admit is almost entirely true—almost. Ecosee is related and dependent upon all of these various eco-studies but is the next logical extension in a discursive environment to Guy Debord's "society of the spectacle." Part of this project arises from what W. J. T. Mitchell (1994, 11) calls "the pictorial turn," where images are becoming more of a problem for public discourse. Besides Debord, who identified the problem of the society of the spectacle, ecosee invests heavily in the promise of electracy, as invented (and termed) by Gregory L. Ulmer (2003). Ulmer, through grammatology, sees the apparatus of literacy failing, because it does not support the technology of the digital Internet, which relies heavily upon the category of the image, a category for which we have developed no logic. One of the purposes of the humanities is that we teach people to become citizens in a democracy that relies upon literacy. Part of electracy (which is to the Internet what literacy is to print) is to help citizens think with the image. Ecosee contributes to this effort in respect to nature, if not also allowing readers/viewers to see systems of image at work rather than working alone.

The rhetorical constructs in *Ecospeak* and other investigations of environmental rhetoric are important in understanding what role the image might have, but while ecosee is motivated from work within environmental studies, the visual aspects are advanced by another realm of research. It traces its roots to other scholars and writers such as previous works by *Ecosee* contributor Cary Wolfe, whose questions regarding the animal other and the idea of being human force us to rethink the very image of animal and theorize the very construction of animal and human. Wolfe's brilliant 2003 *Animal Rites: American Culture, the Discourse of Species, and Posthumanist Theory* and his powerful 2003 collection *Zoontologies: The Question of the Animal*, both of which take into account the "question of the animal" by way of critical theory and theorists, as well as his 1998 *Critical Environments: Postmodern Theory and the Pragmatics of the "Outside,"* a remarkable book of critical theory, stand as central in motivating our work toward a concept of ecosee. As we mentioned earlier, Steve Baker, whose insightful and moving examination of the imagery of animals as it has been employed in performance, theory, and philosophy, also provides ground from which ecosee departs. Other projects in the postmodern disruption and critique of traditions of understanding nature, ecology, science, and other similarly politically loaded terms have encouraged us to pursue this project: Donna Haraway's ongoing work in the philosophy of science and feminisms, taken up in *Simians, Cyborgs, and Women: The Reinvention of Nature* (1990); *Primate Visions: Gender, Race, and Nature in the World of Modern Science* (1990); *The Companion Species Manifesto: Dogs, People, and Significant Otherness* (2003); *Modest Witness@Second_Millennium. FemaleMan Meets OncoMouse: Feminism and Technoscience* (1997), and *When Species Meet* (2007); Bruno Latour's *Politics of Nature: How to Bring the Sciences into Democracy* (2004); Kate Soper's *What Is Nature?* (1995); and Sean Cubitt's *Eco Media* (2005).

Likewise, W. J. T. Mitchell's groundbreaking work in *Picture Theory* (1994) provides a useful, critical eye for looking at pictures and is central to the development of ecosee. Specifically, Mitchell's analysis of the relationship between word and image provides a starting point for understanding the interaction between ecospeak and ecosee, which we might correlate to verbal and visual theories of environmental discourses. Mitchell claims that he does not want to develop a "picture theory" so much as "to picture theory as a practical activity in the formation of representations" (6). Similarly, ecosee functions not just as a nominative term but as a verb, a way of seeing ecologically. One who ecosees looks at images not just for their environmental focus and how they represent the environment but also how that image fits into the larger ecosystem of images and texts. Ecosee asks how an image interacts with other images and texts, how it shapes them, and how it is shaped by them.[3]

While we might try to understand images alone, that is, without attaching to them an external language that exists outside of the image frame, to do so would be problematic and might also be unethical. Images rarely occur without any connection to text, and practical experience tells us that within our culture of communication, one must understand both media to make sense of the constant images that clamor for attention. The category of the image inherent in electracy does not replace literate categories but supplements them, just as Walter Ong demonstrates that the apparatus of literacy does not wholly replace orality. In writing *Picture Theory*, Mitchell explains that

> one polemical claim of Picture Theory is that the interaction of pictures and texts is constitutive of representation as such: all media are mixed media, and all representations are heterogeneous; there are no "purely" visual or verbal arts, though the impulse to purify media is one of the central utopian gestures of modernism. (1994, 5)

Mitchell points out the relationship that images and text have, the "sisterhood" that binds them as familial. This relationship extends to theories of ecosee, where we must understand both how images of environments work and the lingual "messages" that might lie behind those images. Given a postmodern world where media mix and become heterogeneous representations, we might also look at this world in terms of Jean Baudrillard's (1994) theories of hyperreality and recognize that we might not be seeing what we are really seeing.[4] In defense of his work, Mitchell goes on to claim that

> for anyone who is skeptical about the need for/to picture theory, I simply ask them to reflect on the commonplace notion that we live in a culture of images, a society of the spectacle, a world of semblances and simulacra. We are surrounded by pictures; we have an abundance of theories about them, but it doesn't seem to do us any good. Knowing what pictures are doing, understanding them, doesn't seem necessarily to give us power over them. (1994, 6)

But James Elkins suggests that seeing is not simply a passive activity, and that we do have the power to change perhaps not the image itself but what that image constructs. Referring to those "take only photographs" signs we mentioned earlier, Elkins suggests that these instructions obscure the activity of seeing:

> In the national parks there are signs reading, "Don't take any-thing but photographs." It is true that the landscape suffers only infinitesimal change when it loans me a few photons. But we mistake that for the nature of seeing. I may not change a pine tree by taking its picture, though I obviously do affect a bison or a bear by taking its picture. Some national parks have problems with tourists who lure bears with food in order to take their pictures. (And this is where there is truth in that phrase, "taking a picture.") Years ago in Yellowstone I saw a group of cars parked by the side of the road. People were standing at the roadside with their binoculars, looking out across a wide val-ley. When I got out my binoculars I could see what they were watching: in the far distance a man with a camera was running full-tilt after a bison. I doubt Yellowstone has any problem with people mobbing pine trees or patches of turf. What the tourists see is driven by their desire: on the one hand they want large animals, dangerous scenes, and close encounters with white fangs, and on the other they want bucolic, sublime, and picturesque landscapes. Wildness and wilderness are the two goals, and there is very little seeing of botany, geology, miscellaneous zoology, or unpicturesque landscape. Most of Yellowstone is invisible, even though it is there to be seen. (1996, 33)

We never see the whole picture, but what we see is always motivated by desire, what we want to consume as image. Looking is not the passive process of photons penetrating our pupils and reflecting upon the retina, and neither does the action of seeing simply consist of the motion of our eyes. These images, says Elkins, "are not just passively recorded in my mind. Looking immediately activates desire, possession, violence, displeasure, pain, force, ambition, power, obligation, gratitude, longing" (1996, 31).

This brings up the question of ethics and why a study of ecosee is so necessary. Even if we can claim to understand the literate aspects of environmental rhetoric or discourse, we do not yet understand how images contribute to this discourse. Sidney I. Dobrin and Christian R. Weisser claim in *Natural Discourse* (2002) that there is no nature, that

humans always construct it through discourse. So Elkins claims: "Sometimes the desire to possess what is seen is so intense that vision reaches
outward and creates the objects themselves . . . if the desire grows large
enough, it can impel us to make what we want to see out of whole
cloth" (1996, 29). Whether we want a pristine coral reef, cuddly bears
selling Hummers, oil fields coexisting with Alaskan caribou, or animals
applauding General Electric, any visual argument can be made out of
the visual cloth that is the environment, because in the end the environment is just another image to be taken. This does not mean that other
materials are not physically taken from the environment, but that this
kind of taking is predicated on the taking involved in the desire that
accompanies seeing in the first place.

As we hope this collection makes clear, ecosee is not just a phenomenon of visual rhetoric that exists out in the world but is also a way of
seeing. But since ecosee looks for rhetoric in the visual, it does not do
so for purely hermeneutic reasons but also heuretic ones. If activism is
inherent in any environmentally charged mode of inquiry, then the activist using ecosee asks not just what an image means but how one can use
its rhetoric and composition techniques in order to construct one's own
images. In this way, ecosee shares much with ecocomposition, and the latter
should include how ecosystems of writing also include images and, necessarily, how to write images within writing environments. We say "write"
images here, because if we follow the grammatological argument made by
Ulmer, then ecosee is already ecocomposition, since writing images is the
next step in the evolving language apparatus. If, as Elkins shows, seeing is
an (act)ivity, then there is hope that it can lead to new kinds of activism,
ones that are supported by the Internet and Debord's spectacle.

The chapters that make up *Ecosee: Image, Rhetoric, Nature* are organized into four parts in an attempt to bring together similar positions,
arguments, and issues. We would like to think that the organization of
this book emerged organically, providing a logical navigation through the
pieces, but it did not.[5] Of course, this organization is artificial, used for
convenience; the chapters themselves are more dynamic than the organization suggests, more sophisticated than the rubric into which we have
forced them. The relationships between the chapters, the possibilities of
what they suggest, and the work begun by the contributors toward ecosee
require that we look beyond this rubric to other textual ecologies. To
limit reading these contributors' work to the framework imposed does a
disservice to the possibilities of what they present here, and we do not
mean to limit their possibilities through this organization.

Part 1, "How We See," brings together five chapters that initiate
our conversation of ecosee. In "A Rhetorical Look at Ecosee," Sean
Morey addresses some of the rhetorical features that ecosee shares with

environmental rhetoric, specifically that discussed by Killingsworth and Palmer in *Ecospeak*. However, considering scholars such as Gregory L. Ulmer, Roland Barthes, and W. J. T. Mitchell, Morey suggests that environmental images have their own logic, and that a visual rhetoric of ecosee cannot depend upon traditional notions of rhetoric in order to explain it. Ultimately, in order for any debate to occur through ecosee, Morey explains, we must not only be able to read such images about the environment, but we must also be able to make (and teach to make) these images as well.

Bart H. Welling's provocative and thoughtful "Ecoporn: On the Limits of Visualizing the Nonhuman" examines ecopornography, a concept that describes nature-centered photography as having parallels to human-based pornography. Through his article, Welling expands upon the concept of ecoporn, noting that "ecoporn—*as*—porn places the viewer in the same asymmetrical, sexualized relationship to its subjects as standard pornography, even if its primary goal is not sexual arousal." Ultimately, Welling argues that environmentalists need to rethink the human place in its relationship to these nonhuman subjects and develop new visual practices that break out of the commercializing, anthropocentric goals of ecopornography and can help us think up new ways of seeing a nature that "looks back."

In the insightful "Ecology, Images, and Scripto-Visual Rhetoric," Heather Dawkins makes the case that art historians, who usually focus on fine and experimental art that is considered antirhetorical, should also analyze conventional images that often function as a means of persuasive communication. However, much analysis that art historians do overlaps with rhetorical studies. As an example of such rhetorical readings, Dawkins examines how images function within the environmental rhetoric of the 2005 Greenpeace calendar, the pictorial book *Massive Change*, and how both differ from images in Rachel Carson's *Silent Spring*. She concludes that the meaning of these images is produced by the interaction of image and text, what she calls a "scripto-visual matrix."

Spencer Schaffner's "Field Guides to Birds: Images and Image/Text Positioned as Reference" looks at the visual construction of nature by examining field guides to birds. In his intriguing chapter, Schaffner outlines the discrete visual elements that contribute to the distinct forms of visual classificatory discourse in contemporary birding field guides. While field guides are usually thought of as reference material, Schaffner explains that they present images that are not just representations of birds but create taxonomic expectations for the bird watcher that have an impact on how the watcher understands nature. Field guides to birds, Schaffner explains, provide a "taxonomic authority" that produces "specific ways of considering and visualizing the environment."

In the illuminating chapter "Eduardo Kac: Networks as Medium and Trope," Simone Osthoff examines the work of Eduardo Kac, offering a brief overview of his art in general as well as focusing on specific works: *Rara Avis*, *Time Capsule*, and *Rabbit Remix*. She argues that Kac's work influences both the understanding of the natural environment and the "environment of art," and that not only does Kac's art exist in a network and create a network, but also that his theoretical essays "constitute an intrinsic part of his networked ecology," showing the interrelationship between image and text.

Part 2, "Seeing Animals," begins with Cary Wolfe's remarkable chapter, "From Dead Meat to Glow-in-the-Dark Bunnies: Seeing 'the Animal Question' in Contemporary Art," in which he explores two questions: one about the ethical standing of nonhuman animals, and the other about the difference that a particular artistic strategy makes for representing these animals. In this chapter—tied directly to his ongoing projects that pose the "question of the animal"—Wolfe explores these two questions primarily by contrasting the work of two artists, Sue Coe, whose work *Dead Meat* depicts various scenes of animals slaughtered in factory farms, and Eduardo Kac, specifically his works *The Eighth Day* and his transgenic art such as *GFP Bunny* (a transgenic rabbit that glows under ultraviolet light). Wolfe compares what these artists bring to the viewer and argues that while they both offer posthumanist understandings of nonhuman animals, they do so in very different ways, and with different effects.

Following on the heels of Wolfe, Steve Baker's intriguing chapter, "'They're There, and That's How We're Seeing It': Olly and Suzi in the Antarctic" furthers the work he began in *The Postmodern Animal*, addressing the works of Olly and Suzi, British artists who must go into the environments of the nonhuman animal subjects they depict in order to artistically represent them. In his chapter, Baker focuses on Olly and Suzi drawing leopard seals in the Antarctic and argues that over the message or intention produced by a photograph, art can only add particularities, not generalities. Olly and Suzi have to experience the environment of their subject, for it is only in this particularity that it can be understood, and this understanding itself is particular, an understanding that is "how they're seeing it." Their art, by providing this particular, offers a disruption of the general way that humans look at animals.

Part 2 concludes with Eleanor Morgan's "Connecting with Animals: The Aquarium and the Dreamer Fish," in which she makes the important argument that to look at nature is to get caught up in a system of scientific production, mythical production, and material production. Observing the natural, she argues, transforms it. Moving from the capture of the rare dreamer fish to its storage in the Royal British Columbia Museum, Morgan asks the vital question, "How do we look at nature?" She reflects on the inherent dangers of looking, and in doing so, she works to develop a theory that casts animals not as objects of our looking but as activity.

Part 3, "Seeing Landscapes and Seascapes," opens with Pat Brereton's inventive chapter "Farming on Irish Film: An Ecological Reading," in which Brereton, working from methods he developed in his earlier book *Hollywood Utopia*, provides a close reading of three Irish films—*The Field*, *The Secret of Roan Inish*, and *How Harry Became a Tree*—in order to argue that ecology has become a "new, all-inclusive, yet contradictory meta-narrative," that has been present in mainstream film since the 1950s. Brereton's examination of Irish film works to the end of exciting an awareness of interdependence with environment that is visually manifest in film media.

Teresa E. P. Delfín's "Postcards from the Andes: Politics of Representation in a Reimagined Perú" perceptively argues that visual media that portray nature often have as much to say about what nature is as what it is not; visual media work to create a pleasant disorientation between that which is in the frame and what is immediately outside it. But these generalizations cease to hold true, Delfín contends, in cases of third world visual representations of nature. Rather than creating a case for its own difference, images of nature from underdeveloped regions often appear limitless, regardless of the physical imposition of frames, borders, and edges. Third world landscape photographs also are frequently contextualized or captioned to appear normal or native—an everyday part of a context of underdevelopment. This is nowhere truer than in the case of postcards, Delfín explains. Focusing her study of nature-based Peruvian visual rhetoric with attention to the hegemonic nature of literacy in twentieth-century Peru, Delfín maintains that due to inadequate access to education, coupled with the considerable role that literacy has played in the continued subjugation of Peruvian *campesinos,* writing has been inaccessible as a technology for peasant self-representation. In the absence of a contemporary *campesino* literature, Delfín considers the "rival media" of landscape photography and indigenous portraiture in Peruvian *peasant* self-representation.

Kathryn Ferguson, in her thought-provoking contribution, "That's Not a Reef. Now *That's* a Reef: A Century of (Re)Placing the Great Barrier Reef," examines visual images as supplement to the real vis-à-vis the Great Barrier Reef. Beginning with Saville-Kent's first black-and-white photographs of the reef's creatures sent to London in 1891 and moving through to Digital Dimension's integration of 3-D animated images in their 1999 *Ocean Empires*, Ferguson questions notions of authenticity and mimesis in a historically contextualized century of visual images of the Great Barrier Reef. Limiting her analysis to images that specifically lay claim to accuracy and representations of the "real world," she considers the implications of the fact that these images of the reef always refer to something that has preceded them and are thus never the origin but supplement and exceed the origin in ways that may well survive the origin.

The final part of *Ecosee*, "Seeing in Space and Time," begins with Quinn R. Gorman's "Evading Capture: The Productive Resistance of Photography in Environmental Representation," a detailed consideration of a double bind presented through visual representation. Gorman contends that this double bind offers only two options, both of which result in an undesirable "capture" of the world. We either allow the world to be "captured" by the discourse of realism, Gorman posits, which asserts the competence of representational mimesis to reproduce the world in words, or we allow it to be "captured" by the discourse of textuality, which claims that the interests of a natural Other are inevitably and utterly invaded by our own cultural baggage. Within this context of problematic environmental representation as a whole, Gorman postulates, photography perhaps holds a potential place as the medium that uniquely supplies the ground for an ethics that refuses the very *possibility* of capture.

In "The Test of Time: McLuhan, Space, and the Rise of *Civilization*," Tom Tyler forwards the idea that Marshall McLuhan, the once-heralded "oracle of the electronic age," explored, the social and cultural environments created by media technologies and the modes of perception engendered in those who found themselves immersed in media culture. In this chapter, Tyler makes the powerful argument that digital games produce a form of electronic "acoustic space," an instantaneous, inclusive, decentered environment quite distinct from their carefully realized but ludologically irrelevant backstories. Taking as a case study Sid Meier's *Civilization* series, Tyler examines the involving engagement and awareness that digital games require, as well as the equivocal environmental rhetoric of this enduringly successful title.

In the penultimate contribution, Julie Doyle astutely contends in "Seeing the Climate?: The Problematic Status of Visual Evidence in Climate Change Campaigning" that the effectiveness of visual rhetoric as a persuasive discourse within environmental campaigning reached a crisis point in the history of climate change communication. International environmental groups such as Greenpeace often are dependent upon the photographic image to provide evidence of environmental degradation and threat in order to persuade the public and governments to take action. As a result of this reliance, Doyle argues, efforts over the last decade to bring awareness to a skeptical global audience of the *potential* impacts of human-induced climate change were constrained by the very lack of visual evidence about this issue. This lack calls attention, on the one hand, to the problematics of communicating an "unseen" environmental issue such as climate change within the confines of the visual rhetoric of much environmental discourse. At the same time, she explains, these limitations are inscribed more specifically by those of photography as a discourse of visual evidence and truth, unable to visualize, and thus make "real," future environmental threats. Doyle argues that the history of climate change campaigning underlines the interconnections and constraints of both visual and environmental discourse in the communication of this global concern. The lack of visual evidence for events such as global warming, she explains, reflects broader cultural investment in "seeing" and the visual as a primary form of knowledge, while illustrating the privileged role of the visual within discourses of "nature" and the environment.

Following the selections in the four parts of *Ecosee*, we are privileged to be able to include an Afterword to the collection supplied by M. Jimmie Killingsworth and Jacqueline S. Palmer. This Afterword brings together the contributions of *Ecosee* to consider possibilities regarding what work like this might lend in the future. Turning to their own work in visually representing roadkill, Killingsworth and Palmer offer a contextualization, placing the work of visual representation in ecological relation to verbal communication, writing, history, mythology, and technology.

In looking at environmental images specifically, we hope that others can develop theories about them and at least help us understand how we visually and imagetextually represent nature, places, spaces, and environments. While this may not allow us to change our relationship to the image, to give people a power of the image, it at least provides an opening to begin understanding the role of the visual in the political construction and control over nature. Early in the Introduction to *Ecospeak*, Killingsworth and Palmer claim that their book offers "little more than a point of departure for further research" (1992, 2); this, too, is one such departure.

Notes

1. We use the terms *environmentalist* and *anti-environmentalist* as generalizations for different groups that do not necessarily share the same viewpoints. For example, environmentalists include preservationists and conservationists,

even though the two groups approach environmental activism from different perspectives. While also a generalization, preservationists wish to save nature for its intrinsic value, while conservationists wish to "conserve" nature to make it available for social (human) needs.

2. Of course, this argument is flawed, in that what we really mean to say is that sighted individuals experience nature through seeing. One of the immediately recognizable flaws of ecosee is its failure (not yet) to account for nonvisual images and to address the role of visual arguments for those with sight disabilities. Similarly, ecosee, thus far, fails to address issues of access regarding visually impaired "seers" of nature. Dobrin takes up this issue in *Cracks in the Mirror*.

3. Just as the discourse of ecology provides a tool for scientists to study the relationships in an ecosystem, ecosee provides a tool to understand how images function within an ecos(ee)stem.

4. See Baudrillard, 1994.

5. We wish to thank the reviewer for State University of New York Press who suggested this organizational strategy.

Works Cited

Baker, Steve. 2000. *The Postmodern Animal*. London: Reaktion Books.
———. 2001. *Picturing the Beast: Animals, Identity, and Representation*. Champaign, IL: University of Illinois Press. Reprint.
Baudrillard, Jean. 1994. *Simulacra and Simulation*. Trans. Sheila Faria Glaser. Ann Arbor: University of Michigan Press.
Berger, John. 1977. *Ways of Seeing*. New York: Penguin Books.
———. 1980. *About Looking*. New York: Vintage Books.
Cubitt, Sean. 2005. *Eco Media*. New York: Rodopi.
Debord, Guy. 1995. *The Society of the Spectacle*. Trans. Donald Nicholson-Smith. New York: Zone Books.
Dobrin, Sidney I. *Cracks in the Mirror*. Albany: State University of New York Press. Under contract.
Dobrin, Sidney I., and Christian R. Weisser. 2002. *Natural Discourse: Toward Ecocomposition*. Albany: State University of New York Press.
Elkins, James. 1996. *The Object Stares Back: On the Nature of Seeing*. New York: Harvest.
Gottlieb, Robert. 2001. *Environmentalism Unbound: Exploring New Pathways for Change*. Cambridge: Massachusetts Institute of Technology Press.
Gombrich, E. H. 1960. *Art and Illusion: A Study in the Psychology of Pictorial Representation*. Princeton, NJ: Princeton University Press. Reprint, 2000.
Hajer, Maarten A. 1995. *The Politics of Environmental Discourse: Ecological Modernization and the Policy Process*. Oxford: Oxford University Press.
Haraway, Donna J. 1990. *Primate Visions: Gender, Race, and Nature in the World of Modern Science*. New York: Routledge.

————. 1991. *Simians, Cyborgs, and Women: The Reinvention of Nature.* New York: Routledge.

————. 1997. *Modest Witness@Second Millennium. FemaleMan Meets OncoMouse: Feminism and Technoscience.* New York: Routledge.

————. 2003. *The Companion Species Manifesto: Dogs, People, and Significant Otherness.* Chicago: Prickly Paradigm Press.

————. *When Species Meet.* 2007. Minneapolis: University of Minnesota Press.

Killingsworth, M. Jimmie, and Jacqueline S. Palmer. 1992. *Ecospeak: Rhetoric and Environmental Politics in America.* Carbondale: Southern Illinois University Press.

Latour, Bruno. 2004. *Politics of Nature: How to Bring the Sciences into Democracy.* Cambridge, MA: Harvard University Press.

Lessing, Gotthold Ephraim. 1957. *Laocoon: An Essay upon the Limits of Painting and Poetry.* Trans. Ellen Frothingham. New York: The Noonday Press.

Lyotard, Jean-François. 1988. *The Differend: Phrases in Dispute.* Trans. Georges Van Den Abbeele. Minneapolis: University of Minnesota Press.

Mitchell, W. J. T. 1986. *Iconology: Image, Text, Ideology.* Chicago: University of Chicago Press.

————. 1994. *Picture Theory: Essays on Verbal and Visual Representation.* Chicago: University of Chicago Press.

Plato. "Gorgias." 1989. In *Plato: The Collected Dialogues,* ed. E. Hamilton and H. Cairns, 229–307; trans. W. D. Woodhead. Princeton, NJ: Princeton University Press.

Ruskin, John. 1903–1912. *The Works of John Ruskin.* Ed. E. T. Cook and A. Wedderburn. 39 vols. Vol. 19. London: George Allen.

Soper, Kate. 1995. *What Is Nature?* Oxford: Blackwell.

"Time and Space: The Significance of the Cave." 2004. *The Cave of Chauvet-Pont-d'Arc.* http://www.culture.gouv.fr/culture/arcnat/chauvet/en/index.html, accessed December 14.

Ulmer, Gregory L. 2003. *Internet Invention: From Literacy to Electracy.* New York: Longman.

Walker, Chris. 2005. "Earth from Space." In *Views of the Solar System,* ed. Calvin J. Hamilton. http://www.solarviews.com/eng/earthsp.htm#ref, accessed April 5.

Wolfe, Cary. 1998. *Critical Environments: Postmodern Theory and the Pragmatics of the "Outside."* Minneapolis: University of Minnesota Press.

————, 2003a. *Animal Rites: American Culture, the Discourse of Species, and Posthumanist Theory.* Chicago: University of Chicago Press.

————, ed. 2003b. *Zoontologies: The Question of the Animal.* Minneapolis: University of Minnesota Press.

Part 1

How We See

A Rhetorical Look at Ecosee

Sean Morey

[T]he ozone hole is too social and too narrated to be truly natural; the strategy of industrial firms and heads of state is too full of chemical reactions to be reduced to power and interest; the discourse of the ecosphere is too real and too social to boil down to meaning effects.

—Bruno Latour, *We Have Never Been Modern*

If one considers writing as a system for representing nature, a technology used to signify an outside world, then images are as much a part of that system as words.[1] At the same time, writing is not just a vehicle to this world but its ontology. Raúl Sánchez posits that writing is not (only) such a technology or vehicle but also creates identity, in this case, what we identify as "natural." In *The Function of Theory in Composition Studies*, Sánchez confronts the problem with writing as a phenomenon approached hermeneutically:

In the simplest terms, we can describe the hermeneutic disposition as the steadfast and persistent belief in a consequential difference between words and things. In composition studies, most theoretical work subscribes to this belief and, in turn, to the assumption that writing's most salient feature is its ability to represent *something else*, something that is not itself related fundamentally to writing or language. In contrast, I take representation neither as writing's signature feature, nor as an ontological given (as literary studies seems to do), but as a structural component within a general system of discursive circulation and dissemination. From this perspective, the function of composition studies, and of composition theory in particular, is to describe and explain all features of that general system. (2005, 4)

Images—whether appearing as words, pictures, movies, or any visual medium—are not just hermeneutic stabs at showing nature; instead, images compose nature. In natural history terms, taxonomy creates as much as it names. The scientific name for the permit, a fish in the carangid family, is *Trachinotus falcatus*, the species name *falcatus* meaning "armed with scythes" in Latin. Having caught and released many permit, I can offer that the fish is not "armed" with anything, although the image of its dorsal fin sticking out of the water does look like a sickle. However, we construct this association through discourse, using words not to represent the permit's dorsal fin but to generate the association between image and word. Such a theory accords with those of Sidney I. Dobrin and Christian R. Weisser, who write that we construct nature through discourse rather than represent or explain it.[2]

With whichever one of these you agree (writing as technology or generator), as Latour (1993) and Sánchez (2005) suggest, words and the environment operate within a system, and it is difficult to place a border around a system. So just as one cannot separate words and nature from this system, neither can one remove images. They have their own specific niche in the mediascape. As E. H. Gombrich posits:

> Biologists use the term *ecological niche* to describe the environment that favours a particular species of plant or animal. What is characteristic about these situations is again the constant interaction between the factors involved. The rainforests of Brazil could only have developed in a tropical climate, but they are known to influence the climate in their turn. (1999, 10)

We do write and construct understandings about the environment through words, but ours is not (only) a literate society. If we exist, as Guy Debord argues, within a society of the spectacle, then our current notion(s) of environment and nature could only have developed within a culture of seeing and understanding nature in terms of images; and this understanding, a construction of nature through images, has direct material effects on how we treat nature.

Latour, in describing the difficulty in putting a "natural" border around the ozone layer, or a "social" border around corporate interests, illustrates not only a discursive construction of nature but also the natural construction of discourse. A system of nature influences our words for that nature, but these words then create what we see. That is, nature and language are too intertwined to be taken apart, and so image and language are too intertwined to be taken out of the system to understand what they "mean," even though they might have meanings that influence "physical" reality. But just as nature is "real" and operates outside of language use, we do not often acknowledge that images also operate outside of the apparatus of literacy, where words fail to make sense.

This failure of the literate apparatus takes nature into the project of Gregory L. Ulmer, who argues that the category of the image, spread through the digital Internet and the society of the spectacle, operates according to a technology that is not supported by literacy but rather what he terms *electracy*.[3] Ulmer argues that we must develop a logic of the image, specifically because images have effects that words do not, effects that have "real" ramifications; we have to pay attention to images because they short-circuit critical reason and influence how people behave toward each other and toward the earth, behavior that receives little attention. Ulmer writes that the photograph, or image, can "stimulate involuntary personal memory" (2003, 44) in a way that written text cannot and thus bypasses the literate ability to critically interpret the image because it plays on emotion rather than reason. This involuntary response most often happens from what Roland Barthes calls the third meaning of the image, a level of meaning that Ulmer states "the literate apparatus was not suited to exploit . . . fully" (45).[4]

Part of the project of ecosee is to work toward this image logic, toward how the environmental image becomes a category for electrate thinking. The project as a whole requires the participation of many disciplines, from which we can and should study images from their variety of perspectives, ranging from art history and new media to so-called hard sciences such as physics and ecology. I hope that the varied backgrounds of this volume's contributors reflect a constructive tension that helps theories of ecosee adapt to the visual environments of images, natural or otherwise. But from the perspective of a scholar situated in rhetoric and composition, what I offer here is not so much a "grounding" of ecosee in any foundational sense (as if theory needs some sort of hegemonic discursive mode upon which to germinate) but more of a contextualization within this perspective, trying to join the grammatological gap between writing words and writing images. As W. J. T. Mitchell explains, we rely upon ekphrasis to (literately) understand images, and since our understanding of images is grounded (for now) in their description through language, I want to compare the rhetorical features of the environmental image with those of the environmental word, specifically those of M. Jimmie Killingsworth and Jacqueline S. Palmer.

Many of Killingsworth's and Palmer's insights into the rhetoric of ecospeak lend themselves to a study of the visual rhetoric of ecosee.[5] I do not mean to suggest that theories of ecosee must necessarily evolve out of written and spoken discourse, but that Killingsworth's and Palmer's theories of ecospeak, and theories of environmental rhetoric, such as those from Hajer, provide a transition to a study of the rhetorical devices used in ecosee. Ecosee shares many of the rhetorical features of ecospeak. The same groups, from political action groups to the media, use similar methods to influence a viewpoint of nature. From a basic rhetorical standpoint, ecosee uses ethos, pathos, and logos in similar ways; consider the use of ethos in an appeal from Sylvia Earle. While Earle has campaigned through much of her career to stop offshore oil drilling in many parts of the world, in a television advertisement for the oil company Kerr-McGee she now informs television viewers that it can be done safely rather than not at all; in their attempts to stop factory farming, consider the use of pathos in films such as *A Peaceable Kingdom*; consider the rhetoric of science as an appeal to logos, as in documentaries on the Discovery Channel that depict experts explaining the relative risk of shark attacks. These are basic examples, but in many ways the rhetoric of the visual elements of environmental discourse, whether conscious or not of being eco-political arguments, works similarly to written discourse about such issues.

Like ecospeak, ecosee functions on the local level, using images of local concerns to communicate environmental issues. Artists create images along parts of Big Pine Key, Florida, that portray friendly looking endangered Key deer, hoping these images that recall "Bambi" might convince drivers that they should slow down so as not to cause another roadkill. Along the same highway are representations of trophy fish hanging from marinas—giant fiberglass mako sharks and blue marlin—that advertise the charter fishing industry, showing the abundance of the ocean, the spoils to be taken, and that this environment is open for business, although it is "open" because some prior authority has deemed these fish available for the killing, a commodity to be "taken," bound to a culture that says it is okay to kill and hang fish because of the position they occupy/are granted/are regulated to/forced into by this authority. Like the bioluminescent appendage of an anglerfish, these "fake" fish lure in the passersby—drawing them into the marina—where they can then see real fish from the day's catch being displayed in front of the charter boats, yet another lure to bring them in. Traveling along this stretch of U.S. 1, one cannot drive across a single mile marker of the highway without seeing a message about the environment represented in the image. Moreover, one cannot drive across U.S. 1 without seeing the environment either alongside the road as mangrove trees or in the water below the bridges as a visual product where one can sight fish, scuba dive, or sightsee through glass bottom boat rides. This environment is an expected environment, one based upon images of "paradise" already presented by tourism advertisements and images of tropical locations. Even though the Florida Keys is not the same hyperreality these images portray, it is expected (literally, looked out for). As Joy Williams writes, "Nature has become simply a visual form of entertainment, and it had better look snappy" (2001, 12). When we see something that is not expected, then this disjoints our sense of the visual. Outside of the context of a commercialism naturalized along U.S. 1, the same fiberglass mako shark would look ridiculous hung along Massachusetts Street of Lawrence, Kansas. The rhetorical situation no longer serves one of advertising but that of a joke. And an image of that same fish in a seafood restaurant in Denver means something else again: this is not a place to catch mako sharks, but it is still a place where they may be "taken." Thus to function rhetorically, these images depend upon a certain rhetorical context. But more than just an "idea" of the Keys, a literate concept of the Keys' essence, these images create a mood about the Keys, one associated with relaxation, escapism, adventure, and so on. This mood is not a literate category, but electrate, and electracy requires ecosee to investigate such moods about the environment.

These images also pit groups against each other, and all of the groups along Killingsworth's and Palmer's continuum find some way to use images of nature. The Sierra Club adopts the Sierra Nevada skyline; energy companies use environmentalists such as Sylvia Earle to promote their companies and adopt images of nature, such as seagulls in flight above pristine coral reefs; political campaigns use wolves as analogs for terrorists; and as mentioned by Robert F. Kennedy Jr., government administrations use language such as "Clean Skies" (2004, 3–4). These groups co-opt images of nature to conflate their own image with nature itself or portray nature negatively in order to associate it with something they want the audience to associate with negatively.

Instead of verbal clichés such as "tree hugger," standard images become visual clichés about the environment. Images of mutated frogs with multiple limbs have become symbols of environmental poisoning in general, not just in places where frogs live. The image of the manatee represents not only the concern for other manatees but for the entire ecosystem that they inhabit, if not for the entire state of Florida. These images become familiar and standardized in such a way that they become metaphors for environmental problems, just as Hajer describes the process in environmental discourse. The image of a "tree hugger" becomes a visual metaphor for the larger story of deforestation. Such images work because they invoke a predictable response from public opinion, a response measured and tested through focus groups and public relations firms. Thus a single picture can invoke an entire story, and the media, just as they do ecospeak, reuse these images in a way that spreads ecosee. The media and other creators of environmental images create larger story lines, human interest stories, about the environment. But these stories do not just occur in the print of newspapers; they also appear in the images of television, movies, and the Internet.

As mentioned in the Introduction to this volume, Hajer claims that no other image has dominated environmental discussions more than the photo of planet Earth from outer space. As Killingsworth and Palmer illustrate, *Time* magazine uses a similar image on the cover of its January 2, 1989, edition, a photograph that illustrates how a single image can create a story line that has dominated environmental politics for the last fifteen years. This edition of *Time* featured its "Person of the Year" for 1988, but in place of a person, the usual focus of the annual issue, *Time* named the "Endangered Earth" "Planet of the Year." The cover photo portrays a beaten "globe wrapped in polyethylene and rag rope" (1992, 157). This image helped spawn a renewed interest in environmental concerns and showed a shift in the environmental viewpoint of *Time* magazine. According to Killingsworth and Palmer, *Time* "represented a more conservative political constituency and agenda in the early 1960s" and "sharply criticized [Rachel] Carson's rhetoric in its 'Science' section" (1992, 72). But beginning in the late 1980s, "*Time* is doing more than merely reporting the facts; it has taken up overt efforts to influence future actions," and "the focus of the whole 'Planet of the Year' issue is one action" (1992, 158). *Time*, as a magazine devoted to both written and photo modes of journalism, understands the impact of a picture. The use of a picture that resembles the Apollo photograph is no accident, and the stylization to show the Earth's feeble health is a clever visual rhetorical construct. This edition of *Time* includes environmental issues on a variety of environmental topics, creating a whole story line with one issue, all connected to the cover image. *Time* takes an image, inserts it into the environmental debate, and shakes the expectations of ecospeak: "Until the summer of 1988 . . . *Time* remained true to the categories of ecospeak" (1992, 152). This created a new story line connected to Earth, perhaps the greatest interest a human should have in anything:

> The human interest slant of the magazine . . . has been extended to a "whole earth interest" . . . by 1989, nature had caught the magazine's interest with a crisis-level insistence. As the mythic personification of the earth as the goddess Gaia, *Time* featured its home planet in a position normally reserved for human subjects. (1992, 157)

Although this issue of *Time* is full of environmental articles, the cover photo impacts audiences and shows how an image can disturb the normal pathways of ecospeak, and how environmental images and environmental discourse are connected in an ecology of rhetoric. *Time's* change of perspective also shows the change in Killingsworth's and Palmer's continuum that the media can make, where this "personification of the planet represents a significant step toward the rights-of-nature approach of deep ecologists" (1992, 157) and how a single image can represent that change.

But unlike ecospeak, which is somewhat limited to political discussions, ecosee is more pervasive, because its images spread in a more ubiquitous manner, such as on the cover of a magazine (which itself is further covered as news). Often the image's environmental stance is hidden or implied, or it becomes a mask for other arguments that the image seeks to make. An advertisement for a chicken sandwich may seem only a marketing device, but behind that device hides an eco-political argument or an "educational" message. Consider David Orr's suggestion that "all education is environmental education":

> By what is included or excluded we teach students that they are part of or apart from the natural world. To teach economics, for example, without reference to the laws of thermodynamics or ecology is to teach a fundamentally important ecological lesson: that physics and ecology have nothing to do with the economy. It just happens to be dead wrong. (12)

By similar means, all images are environmental images. Chick-fil-A® advertisements that depict cows holding signs that read "Eat Mor Chikin" portray chickens as not belonging to "nature" proper but existing within a domesticated group of animals specifically marked for human consumption. Although Chick-fil-A® did not invent the chicken ("just the chicken sandwich"), it depicts the proper niche of the chicken (in-between a bun). In this way, such images present a perception of the world and show what is and is not "nature," or how and in what ways we should use nature.

Although I differentiate between ecospeak and ecosee for identification purposes, the two work together in a symbiotic relationship. W. J. T. Mitchell (1994) claims in his book *Picture Theory* that poetry and painting share a "sisterhood"—ecospeak and ecosee share a similar relationship. As an imagetext, the two may inform each other, yet they may also contaminate each other, with ecospeak forcing a discursive construction onto an image, or ecosee inappropriately infecting a discursive context. However, this imagetext must remain, since we can only (literately) understand nature through language; thus given a still dominant literate apparatus, we mainly understand visual representations of nature through ekphrasis. Ecosee thinks not only about image and nature but also about the (inter)play between images and text, the interplay between how physical/imaginary/virtual/hyper environments shape the production of images, and how these images interact with each other in the sphere of visual discourse about the environment. Images (and texts) connect in the same way that the components of an ecosystem connect, and the way that ecosystems connect to form larger environments. Ecosee is a study of the ecology of images in a particular region and the ecology of images in the global infosphere. Ecosee looks at the use of images across platforms and agendas, from politics to advertising to education to entertainment, all of which make up a mediascape of images. It also takes into account that the use of images functions differently across cultures and countries, but that these images come into contact in a larger ecosystem of images and affect the way people view nature. However, the hermeneutic aspects of ecosee are only half of its potential. Ecosee asks not only what images "say" but also how they are made (i.e., the heuretic question). By understanding how an environmental discourse of images is constructed—just as Killingsworth and Palmer show how ecospeak is made—an examination of the nuts and bolts of ecosee allows us to learn how to make, and in turn teach others to make, images that contribute to a visual environmental rhetoric. This does not mean that such rhetoric must adhere to a present understanding of ecosee but instead will change by allowing others to engage with the environmental image and participate within the public exchange about how we see the environment.

Some Semiotics of Ecosee

But the relation of language to painting is an infinite relation. It is not that words are imperfect, or that, when confronted by the visible, they prove insuperably inadequate. Neither can be reduced to the other's terms: it is in vain that we say what we see; what we see never resides in what we say. And it is in vain that we attempt to show, by the use of images, metaphors, or similes, what we are saying; the space where they achieve their splendour is not that deployed by our eyes but that defined by the sequential elements of syntax. (Foucault, 1973, 9)

At a literate level of understanding, images function according to a visual syntax, a visual coding system that can produce meaning. This is not to say that this coding is the only meaning, or even the preferred meaning; other modes of signification, such as Roland Barthes's "obtuse meaning," exist at a level beyond verbal discourse.[6] But even within a system of codes for the image, not every element of nature becomes coded in the visual. It would be nearly impossible to saturate the visual landscape with images of every type of flora, fauna, vista, and environment; instead, ecosee relies upon a limited number of image types to express its range of arguments. Given the diversity of species on the planet, it would be impractical (though conceivably not impossible) to use each and every animal, plant, and mineral as a distinct and meaningful sign. Even the Chinese, who use thousands of ideograms as a written language, traditionally will employ only certain "important" plants and animals within their brush paintings, creating a hierarchy within their natural environment. They historically relied on plants such as the orchid, chrysanthemum, and bamboo. Chinese brush painters Wang Jia Nan and Cai Xiaoli explain that the bamboo

> . . . has been a leading subject of Chinese painting for centuries, and it is also a good example of the way in which painting subjects are more than the representation of a visual or imaginary world. There are more than 280 different sorts of bamboo in China. It comes in a variety of colors, shapes, and sizes, it flourishes in all types of soil. Its parts are used for building, utensils, papermaking, and eating. (2003, 46)

Here we can see how the Chinese view their particular culture and thus determine an appropriate sign for signifying the use of nature and the personal qualities that the Chinese attach to the bamboo, such as endurance and steadfastness. The Chinese also paint the plum blossom, whose name (*li*) refers to both the plant and ethics. By classifying the world with images, the Chinese created a vocabulary of images that conveys ideas about the environment and their relationship to it.

Of course, a country that has no bamboo will use a different tree as a symbol for strength and endurance, if those qualities are important to the cultures that inhabit the region. The Chinese are useful to look at because they developed an image-logic system based upon nature. Since theirs was an agrarian society, they studied nature in order to survive. However, instead of focusing on the essence of individual components of an ecosystem, the Chinese understood these components through their relationships with other elements. By portraying certain elements in a landscape painting, the artist could suggest something about an individual, for all of the relationships that occurred in nature at the macroscopic level occurred for each person at the microscopic level. Thus if one wanted to understand *li*, then one would study *li*, because the relationship that the plum blossom has to the rest of nature can tell the individual something about ethical behavior. However, the portrayal of any one element in a landscape painting is unimportant; what matters is how one subject interacts with all of the other elements in a painting. The artist must not paint a thing but a relationship.

The Chinese perspective of landscapes seems an especially appropriate comparison for a study of visual semiotics. Fernande Saint-Martin discusses that visual language must be analyzed according to the scientific methodology of "analysis." Saint-Martin explains:

> This analytical method aims at bringing to light the interrelations between elements—rather than their hypothetical essence—in the totality that is the visual work. Whether it concerns a painting, a sculpture, a photograph, or an architectural edifice, the work considered as a totality "does not consist of things but of relationships," as Hjelmslev has already proposed for the analysis of verbal language. (1990, 183)

Like a landscape painting, or an environment, the individual elements may be important, but mainly in how they relate to other parts of the landscape. In Chinese painting, one cannot take away the river without drastically altering the "meaning" of the painting and destroying the overall sense of *shi*. Ecologically, the extinction of a keystone predator will dramatically change the remaining relationships between animals, irrevocably changing that ecosystem's ecological destiny. The genre of landscape painting can provide a useful analogy to how we view environmental images and how they combine and constitute a landscape or ecosystem.

Despite the cultural relevancy of certain images, mass media have exported different environmental images around the globe, so that an animal in China, such as the panda, has become an important eco-political image for the whole world. Such images become either ecotypes or "econs."[7] Ecotypes resemble archetypes, categories of animals that may look different from image to image, and may even be different animals within paintings, but serve the same function within those paintings. For example, images of humpback whales or elephants could be used to portray the concept of an endangered species. Ecotypes may represent stock definitions from ecology, and science often uses them as examples, such as predator, prey, consumer, producer, parasite, host, terrestrial, and aquatic. Thus any animal can fulfill one of these roles and serve as an ecotype in doing so. Econs, such as the panda, provide instant associations with organizational groups (such as the World Wildlife Fund), ideas, or movements. Icons from other contexts may even be appropriated by environmental (or anti-environmental) agents and become an econ.

Ecosee relies on a limited number of animal types to portray nature, and these types become archetypes, or ecotypes, that can represent similar populations of animals, or even all of nature. But reducing the available signs from millions of species to perhaps only thousands (or hundreds) is necessary and even proves useful. In *Postmodern Ecology*, Daniel R. White explains:

> Building on Levi-Strauss's observation that a system is made numerically poorer but logically richer by subtraction of elements to make the remainder discrete, Wilden argues, "the point is, of course, that only systems of discrete components are available to COMBINATION and permutation, that is to say, only such systems can properly be said to have anything equivalent to SYNTAX." (1998, 159)

By selecting a small number of the total species, ecosee can take the "differences between animals and make them oppositions." It would not prove useful and effective to create meaningful symbols for three different species of cockroaches, or to create images of every type of shark. Images of the white shark, especially those used to portray it as a "man-eating machine," represent, at least for large segments of the public, all sharks. Such an image provides both an ecotype for the shark, and, in certain cases, it also may present an econ. However, while ecosee might not rely on creating a set of images for every breed of dog, every breed of dog can be portrayed as a "dog," just as every conceivable way to write "dog," or just the letter "d" (d, D, Δ, Ď, d', đ, 𝒟, **d**, D, ad infinitum) can represent that word or letter.

Steve Baker, in his book *Picturing the Beast*, discusses how companies use dogs in advertising, and that ads use stock types (what I

am suggesting can be identified as ecotypes) in order to prey upon
consumers' expectations:

> The dogs most often seen on television—the puppy unrolling
> the Andrex toilet roll, the slow-motion Dulux sheepdog, the
> Crufts champions fed on Pedigree Chum—are a particular kind
> of dog. They are never rottweilers or American pit bull terriers or
> Japanese tosas; they are, in other words, every bit as stereotyped
> and "perfect" as the people in advertisements. (2001, 170)

Even while any breed of dog may portray a dog in general, they are
used as just dogs (albeit "perfect" ones) rather than a particular breed
such as border collies, just as the use of a rat can represent all rodents
or a robin all birds; the emphasis is not placed on their border colli-
ness or robinness, but on their dogness or birdness. However, within
this matrix of animals, types, and referents, certain animals emerge and
become econs, such as the panda or bald eagle.

The bald eagle has served as an econ because of its status as
the American national bird. It becomes the symbol for patriotism or
"America" in many contexts in which it appears, even "objective" nature
documentaries. The major environmental issue that deals directly with
the bald eagle has been its presence on the endangered species list, and
in a way, this presence ironically symbolizes the United States' concern
for environmental issues as a whole. The image of the bald eagle is even
more ironic when energy companies appropriate it into their advertisements. Consider an Americans for Balanced Energy Choices (ABEC)
ad that features a bald eagle with the subtitle "1970." The eagle flies
through a polluted sky, lands, hacks on the sooty air, and proclaims:
"Not a good day for flying." The scene then cuts to the present day,
where the air appears clean, due to clean-burning coal technology
developed in part by ABEC (apparently not due to the efforts of the
Environmental Protection Agency [EPA] and other government-run
environmental regulatory agencies developed post-1970). The nested
symbolism of both "nature" and "America" allows this energy company
to portray itself as both environmental and patriotic (important within
a new rhetoric of patriotism post-9/11).[8] Through an econ, the energy
company can produce the "image" that it cares about America and the
American environment.

Such an appropriation of an environmental image by advertisers
shows how the root for econ (Gk.: *oikos* = home) often can suggest
both ecosystem and economy. Advertising agencies use environmental
images to sell their products or ideas. In Baker's example, a dog food

company will use the "perfect" puppy to persuade a potential customer. In general, this applies to other environmental images, especially well-recognized econs. ABEC does not portray just any bird flying through a smoggy sky but the "perfect" bird. Instead of choosing any bird, it relies on a specific econ to convey certain connotations.

Most ecotypes represent what Baker would call "good animals." He claims, "The image of the bad animal does not exist for advertising," because such an animal would make the product unappealing to consumers. However, I have to disagree with his premise under the adage, "The only good bad animal is a dead bad animal." Advertising often uses bad animals to sell products or services. Consider pest control companies that reinforce the idea that nature is something to be controlled by humans. Advertisements for Terminix® portray pests, usually insectan pests, as forces to be destroyed to enforce cleanliness. By the end of the commercial, these "bad" animals are usually killed and thus become "good." The only "good" living pests are animated, highly anthropomorphized roaches, such as in Raid® commercials, which become comical characters. Yet even these pests are killed by the end of the advertisement, this time by the self-sufficient home owner.[9]

However, beyond establishing a dichotomy between animals that are good and bad, this portrayal of good and bad animals also creates a value judgment of which animals get to count. By constantly referencing a particular species in an image, one excludes other life-forms. Thus focus is given to a particular species continuously, never allowing others to be seen. Ecosee, perhaps unwittingly, creates visual niches for certain animals and competitively excludes others. One rarely sees advertisements promoting the survival of insects unless they are butterflies or some other "cute" and "good" animal. While environmental efforts and commercial interests have readily adopted cetaceans into many ad campaigns, only a few species are portrayed, such as bottlenose dolphins, orcas, humpback whales, and occasionally a blue whale. This excludes other cetacean species, such as the beluga whale, narwhale, sperm whale, and pilot whale. These last two examples are featured more often in news reports and literature. The sperm whale, of course, is the species of Moby Dick, while beached pilot whales frequently appear in news reports, gathering attention from volunteers who try to help them, thus creating a human interest story and operating under the guidelines of ecospeak. However, these two species are poorly represented in images and fail to attain full status as econs or ecotypes.

The image of the manatee is one of the most represented econs in the state of Florida, appearing on billboards, artwork, license plates, marine navigation signs, souvenirs, and the like. The image of the manatee often illustrates the peril its species faces by focusing on its high potential for extinction, or through its use as a marketing device; a business can simply appeal to the audiences who care about manatees, or at least have an interest in them. But images of manatees also become symbolic of the entire coastal ecosystem. The image of the manatee can then serve not just as a ubiquitous econ that suggests one species in peril but as an econ that can represent other endangered species, and also entire ecosystems, families, classes, ideas, or the entire concept of nature, the world, or life. Imagetexts of manatees that include an image of the animal with the text "The Real Florida" say something about how we discursively construct nature, and also about how we econically construct nature—in this case, Florida. One can simply appropriate an econ with established environmental story lines and use that image as a rhetorical device either to lure in customers, to present the illusion that Florida is somehow "natural," or to suggest that the state cares so much about manatees that it invests the idea of its state with the animal's image, even though the state's actual legislative and enforce-ment practices might belie this message. In fact, the quantity of manatee images far outnumbers actual, living manatees, and this preponderance of images suggests not that the species is endangered but quite the opposite: that the population is healthy and thriving. Viewers come to know the manatee through the hyperreal, and this perception suggests hypernumbers of "real" manatees.

The eagle, manatee, humpback whale, and white shark all serve as econs for various purposes. Because they represent patriotism, the Everglades, the ocean deep, or a primordial fear, these animals have symbolic status. However, as we have seen, one can easily appropriate one of these icons, icons that seem to have a fairly static "message," toward many other intended uses. While rodents may appear as "bad" animals in pest commercials, Baker, looking at Mickey Mouse, points out that "just about anything will do as a national animal symbol." Baker goes on to say that "the symbolism itself is seldom very clearly defined, and it is open to manipulation: it is a rough-and-ready symbolism. It is in no way hindered by the fact that its meanings need owe *nothing* to the characteristics of the animals it employs" (2001, 62). Just about any animal can serve as any kind of symbol, and while this may be true as a culture develops, within a temporal "snapshot" of a particular culture, dominant symbolism will be attached to some animals and not others. A study of ecosee should develop a portfolio of these econs, with which one can then compare the different (mis)uses of these animals in visual media. However, the catalogue need not be static but allow for other econs should rhetorical situations demand. One can make visual discourse from such econs, but one also can make new econs when necessary. A heuretic investigation of the econs currently in existence might tell us how to do this. Baker, discussing the specific use of animals as national symbols, sees both positive and negative aspects to such a catalogue:

> With potentially negative connotations crowding in on all sides, the animal looks in retrospect to be among the least secure images for carrying messages about human identity. Given the character of many of those messages, this may be no bad thing. But there is a more positive aspect to the examples. . . . They show nations having chosen . . . to depict not only other or rival nations but also *themselves* in the animal form, or else to define themselves by means of an identification with animals. There is thus a certain equilibrium, or balance of power, in the distribution and operation of these symbols. They serve to remind us that the clichéd notion that our culture always sees animals as inferior need not simply be taken for granted. . . . Even if . . . these supposedly positive animal images have been drained of much of their animality, they are still the culture's chosen iconography. It is too easy to forget this, or else to give insufficient consideration to its significance. (2001, 69–71)

Although Baker is correct in seeing positive aspects in this "balance of power," this equilibrium requires that we question those animals we consider inferior to humans, and then how that becomes displayed within images. If all elements of "nature" are required for its survival and sustainability, then how can any part of that system be "bad"?

This catalogue of econs, this exclusion of some animals in favor of others, points toward both a canon of images that "may" be consumed by the public sphere and also a system of indexicality as it is used in visual semiotics. While the indexicality of signs can be difficult to explain, Keyan G. Tomaselli says it most succinctly:

> The index draws attention to the thing which it refers. For example, a weathercock is an indicator of wind direction. The sign draws attention to the existence of the unseen—it has an existential relationship to the phenomenon it depicts. Wind cannot be seen except in a secondary way through indicators like a weathercock, vane, a wet index finger or some other indicator, like a tree bending in the direction of the airflow. (1996, 30)

These indexical signs allow us to see the unseen. In this sense, we might say that all econs, all ecotypes, display a form of indexicality, since they all refer to the concept of nature. Jean-François Lyotard states, "To every realism, it can be answered that no one can see 'reality' properly called" (1988, 33). I would say that no one can see "nature" properly called, and its existence must constantly be proven through the use of indexical signs, signs that have thus far operated according to the semiotic rules of ecosee. While general rules of visual semiotics may aid us in deciphering the rules of ecosee, the study of the ecology of signs that visually construct nature can also help us determine the specific subset of visual semiotics that pertains precisely to ecosee. As Killingsworth and Palmer explored ecospeak through its specific use of rhetoric, we can explore ecosee through its unique use of visual semiotics.

Ecosee and *Picture Theory*

> But then people have always known, at least since Moses denounced the Golden Calf, that images were dangerous, that they can captivate the onlooker and steal the soul. (Mitchell, 1994, 2)

Since images function within a system of discourse, ecosee cannot be understood only in terms of images but must also be considered in terms of the relationship between image and text—how the two interact with each other by informing, conflicting, and contaminating each other in the Barthesian sense.[10] As I suggested earlier, one method of exploring the logic of environmental images, at least as rhetorical constructs, is through the study of environmental rhetoric and where the two might intersect. W. J. T. Mitchell provides a useful way to understand this relationship, seeing the two not as distinct entities but as related. Mitchell posits that writing and speech might "have the same sort of 'sisterhood' as painting and poetry—a sisterhood of radical inequality, as Lessing and Burke argued—if writing transforms invisible sounds into a visible language" (1994, 116–17). Just as writing and speech and painting and poetry form sisterhoods, so image and text form a sisterhood. The problem with this relationship is that writers often privilege their text to be visual enough without the image. Text itself is an image, but it is often translated back into sound when we read it. And since we are in a culture where images are the dominant visual experience, text must find some other niche in this relationship.

Still, we can understand ecosee and ecospeak as sharing a sisterhood within environmental rhetoric. Ecosee is the dominant form of environmental debate over ecospeak, but people must recognize that this visual debate is taking place, and that it is much more persuasive than any speech by a politician. One way to make this apparent, at least at first, is to describe images to people in familiar terms; one must couch the visual debate within language, and that language is the language of environmental politics. While this may seem limiting at first, viewers must recognize that images are making these same kinds of arguments before they can further understand and accept or reject those arguments. The method that one might use to explain images is through Mitchell's understanding of ekphrasis.

In his most simplistic description of the term, Mitchell defines ekphrasis as "the verbal representation of visual representation" (1994, 152). This is where the verbal sister is "mobilized to put language at the service of vision" (153). Ekphrasis provides the verbal expression of the visual, because the only way we can understand the visual is through language. This makes ekphrasis especially suited for a study of ecosee, for like the visual the only way we can understand "nature" is through discourse.[11] As discussed in the Introduction, not only do we typically engage "nature" through the visual, but we also can use the same method to discursively construct both nature and image. While image and text share a sisterhood, nature and image also share an affinity.[12]

Mitchell complicates the nature of ekphrasis, and we begin to see that it is not just the mere description of an image that occurs but also the site of rhetorical conflict. According to Mitchell, the "central goal of ekphrastic hope might be called 'the overcoming of otherness'" (1994, 156). Language attempts to deal with its visual sibling by discursively constructing it in a certain way that makes this image into language's "own image." However, this leads to the "ekphrastic fear" that a break-down will occur between image and language; images should properly be mute, and language should be without image—the writer should not attempt to produce a painting with words. Mitchell discusses Gotthold Lessing's ekphrastic fear, his

> . . . fear of literary emulation of the visual arts is not only of muteness or loss of eloquence, but of castration. . . . The obverse of ekphrasis, "giving voice to the mute art object," is similarly denounced by Lessing as an invitation to idolatry: "superstition loaded the [statues of] gods with symbols" (that is, with arbitrary, quasi-verbal signs expressing ideas) and made them "objects of worship" rather than what they properly should be—beautiful, mute, spatial objects of visual pleasure. (1994, 155)

This discursive construction of an image, the application of arbitrary signs and meaning onto the visual, corresponds to how humans discursively construct nature. I am not suggesting that "nature" should be a visual site that should be "beautiful, mute," untouched, and available only for viewing, but that other meanings of nature—as a resource, as "wilderness," or as other constructions made throughout history—are as equally problematic as any other act of "idolatry." Just as language attempts to reconcile the "otherness" of the image, language also attempts to deal with the "otherness" of nature. And although nature and the image share a similar relationship to language, we now see that the image attempts to focus on the "otherness" of nature as well.

However, ekphrasis cannot hope to overcome "otherness," unless that otherness is a fixed entity. Mitchell argues in *Picture Theory* and elsewhere that the scientific categories of otherness (symbolic and iconic representations; conventional and natural signs; temporal and spatial modes; visual and aural media) (1994, 156) do not stay within these categories and "are neither stable nor scientific" (157). Better understood as "ideologemes," these categories are "allegories of power and value disguised as a neutral metalanguage. Their engagement with relations of otherness or alterity is, of course, not determined systematically or a priori, but in specific contexts of pragmatic application" (ibid.). As an other, the visual presents a passive site of receptivity, where we may insert our own discourse in order to understand and verbally represent it. "Nature" as "mother" or resource provides this same opportunity as a passive receptacle for language, and images of nature are open to the same practice. The application of ekphrastic practice to either nature or visual representations of nature will depend upon the contexts of how people view nature and their pragmatic goals in discursively constructing nature in a particular way.

The ekphrasis of images, the verbal drawing of images, is where the power struggle occurs in environmental politics, or for that matter in any attempt to assign identity to an image. Since a picture is "silent" and cannot speak for itself, active agents can shape that image's identity and determine for the image what it will "say" to an audience. The verb "say" is an important term, because as Mitchell explains, images become speech acts and can "say" things just as well as words:

> The moral here is that, from a semantic point of view, from the standpoint of referring, expressing intentions and producing effects in a viewer/listener, there is no essential difference between texts and images and thus no gap between the media to be overcome by any special ekphrastic strategies. Language can stand in for depiction and depiction can stand in for language because communicative, expressive acts, narration, argu-

ment, description, exposition and other so-called "speech acts" are not medium-specific, are not "proper" to some medium or other. I can make a promise or threaten with a visual sign as eloquently as with an utterance. While it's true that Western painting isn't generally used to perform these sorts of speech acts, there is no warrant for concluding that they could never do so, or that pictures more generally cannot be used to say just about anything. (1994, 160–61)

Ecosee, and theories of electracy, as developed by Ulmer, would suggest that our culture is moving increasingly toward a rhetoric of images, where most communicative acts occur through visual media. However, those that control what the images "say," the pictorial manipulators that give picture/speech acts their illocutionary force and perlocutionary effects, are hegemonic structures that are able to determine what gets shown. Often the messages given to images cater to our expectations, create expectations, and prey upon these expectations to advance commercial consumption. Thus a hegemony exists that portrays "nature" in a way that becomes accepted and presents expected norms of how nature really "is." These ubiquitous images inundate the perception of nature, found on television and billboards, in movies, on the Internet, in fashion and clothing, or in video games, and they create a hyperrealized state of nature. Moreover, these images are so pervasive and transparent that they are overlooked and numbing, making one less inclined to public or private action.

Theories of ecosee would attempt to make these identities of nature more apparent to audiences who can choose for themselves how they might understand nature, or at least participate in the debate, a participation that includes not just reading images but also producing them. While part of this goal involves using visual semiotics and other logics of the image, another facet involves the rhetorical conventions in which those images work. Although many different rhetorical approaches exist when using images, environmental images typically fall within the rhetorical categories defined by Killingsworth and Palmer in *Ecospeak*. Different groups with various eco-political viewpoints use and/or develop images in advertisements and marketing campaigns that create pairs of opposites, adversaries, in the same way that journalism can write story lines that create characters for environmental debates. In a way, image-texts in general already function according to some of the rhetorical constructs of ecospeak. Image and text become factions that can join to create common messages, or they can work against each other to create conflict. However, while similar rhetorical devices found in *ecospeak* also exist in *ecosee*, the very mechanics of speaking and seeing alter their effects and the interaction between the two communicative partners. In *ecospeak*, two debaters may answer and attack one another with varying levels of discourse, adapt during the communicative event, and perhaps change the other's opinion and advance their agenda. With an environmental image, at least at first glance, little adaptation occurs, and both partners of the "conversation," the image and the viewer, are placed in doubly passive positions. The image has no power to change itself from what it is and can only be seen; the viewer has no way to change the image, though perhaps his or her perspective of it, but even this "seeing" is a passive activity, where the eye simply becomes a receptacle for penetrating light. This is not to say that activity does not occur within the brain upon seeing, and that the mind does not actively construct what it sees, but within an oral debate or a written argument, writers can respond to each other and engage in a discussion. How does one respond to a picture?

At second glance, there are many ways to change such images. The increased use of video sites such as YouTube allows anyone with a video camera and an Internet connection to place her or his own perspective online. Moreover, the community-based/networked nature of sites such as Facebook and Myspace help create communities that can engage in political discussions. I am not saying that this is the primary use of such sites, but such technology does facilitate this. Also, the availability of image manipulation software such as Adobe Photoshop allows users to create and edit images that can be easily captured from the Web. But the image still cannot change itself once put into circulation. And as long as nature is presented in images, the viewer is always the subject, the image always the object, even when images engage in speech acts. This objectivity can easily be transferred to the "real," which is never allowed to achieve a state of subjectivity. Even images that present animals positively still portray them as objects, as animals that need human protection and can never obtain their own agency. While animal activists try to help animals, they only advance their own idea of nature, often a patronizing position of humans as stewards of nature.

This dilemma makes more pressing the ethical question of representation in ecosee. If the viewer of an image cannot change that image, and the image cannot change itself, then theories of ecosee must provide ways to recognize the third meanings of an image and/or allow readers/viewers to determine/construct denotative and connotative meanings for themselves. In writing *Ecospeak*, one of Killingsworth's and Palmer's main goals is to establish a study of rhetoric that is useful for both students of rhetoric and for politically active scientists or other individuals who might wish to alter public discourse through a praxis of ecospeak theory. While I do not feel that theory must necessarily lead to immediate practical use, theories of ecosee should help individuals recognize the conventional rhetorical devices and their intended effects, who can therefore accept or reject those meanings, or, once recognized, construct their own images of nature. As Mitchell puts it, "Perhaps we have moved into an era when the point about pictures is not just to interpret them, but to change them" (1994, 369).

The Ethical Call of Ecosee

> The center post was cut all the way through. The outer posts were each cut more than two thirds through. The great sign rested mostly on its own weight, precariously balanced. . . . Bonnie placed her small brown hands against the lower edge of the sign, above her head, barely within her reach, and leaned. The billboard—some five tons of steel, wood, paint, bolts and nuts—gave a little groan of protest and began to heel over. A rush of air, then the thundering collision of billboard with earth, the boom of metal, the rack and wrench of ruptured bolts, a mushroom cloud of dust, nothing more. The indifferent traffic raced by, unseeing, uncaring, untouched. (Abbey, 1999, 50)

To return to Mitchell's suggestion, how does one change images? This passage from Abbey's *The Monkey Wrench Gang* shows one way: sabotage. In this scene, two of the book's many protagonists, Dr. Sarvis and Bonnie Abbzug, destroy a billboard that says:

MOUNTAIN VIEW RANCHETTE ESTATES
TOMORROW'S NEW WAY OF LIVING TODAY!
Horizon Land & Development Corp.

Within their effort to destroy all billboards, we see a contestation over who gets to occupy the spaces and places called "nature." At the same time that this billboard offers the commodification of seeing mountains, it also occupies a place in the world that has been designated a place for advertising and is itself a thing meant to be seen. Those who control the space to place a billboard are those who get to say, or get to show, what nature is. Dr. Sarvis and Bonnie feel ethically bound to reoccupy this space for what they call "nature," and to exclude billboards from the image of nature that drivers see out of their car windows—to let nature "speak" for itself.

However, while I am not suggesting that anyone change images through illegal destruction, Abbey does point to the problem of (not) seeing ecologically. Although many billboards offer certain eco-political arguments, most people probably drive by them "unseeing, uncaring, untouched." Because these images pervade our roadways, televisions, computers, magazines, books, and newspapers, we become numb and indifferent to the obtuse meanings that they contain. While many obtuse meanings may exist, each image has an obtuse meaning concerned with "nature." Of course, many of these images reflect a preconceived idea of nature as a priori, a "nature" that I suggest, along with theorists of nature such as Sidney I. Dobrin and Christian R. Weisser, does not even exist. Or, as Lyotard asserts, "The ontological argument is false. Nothing can be said about reality that does not presuppose it" (1988, 32).

Through the writings of Jorges Luis Borges, Göran Sonesson discusses the a priori perspective that cultures often assume when picturing nature. Although cultures such as the Chinese depict the bamboo that spreads throughout their region, other cultures overlook such obvious, daily encounters. In his piece "Why There Are No Camels in the Koran: The World Taken for Granted," Sonesson extrapolates on Borges's point that the authors of the Koran felt no need to write about camels:

> The total absence of camels in the Koran shows, according to Borges, the authenticity of the text: although from the horizon of Occidental culture, the camel seems characteristic of the Arabic world, the Arab simply takes it for granted, and so does not bother to mention it. (1989, 30)

From this reading we might ask why the images of nature appear as they do. Which elements of "nature" are not taken for granted and are then overrepresented? If these elements are overrepresented, is this because of concerns about endangered habitats or animals, or because their "place" in nature must be continually reified in order to uphold the normative view of nature? If images of manatees constantly appear throughout Florida, are these images somehow less authentic of the "real Florida," despite their claim to the contrary? Which images are out of view or nonexistent because they do not fit within our perception of "nature"? What, for us, are the camels?

In a very "real" sense, this study of ecosee assumes that it functions within a realm of postmodernism, where little is so certain, and that "nature" as we typically think of it does not exist. That is, we cannot take nature as a priori, but we construct it according to our current understanding of nature, how it operates and how we use it. This might chagrin some scholars, such as Michael E. Soulé and Gary Lease, who write against postmodernizing the environment; but to claim that nature does not exist does not preclude us from studying it or denying that environmental problems exist.[13] Environmental problems stem directly from ecospeak (and ecosee), and thus exist hand in hand with nature as a discursive construction. Global warming is a constructed problem in more ways than one. As Hajer notes, this understanding is based on representations; as I argue, these representations have been largely analyzed through written or spoken discourse, not visual. This needs to change.

The environmental crisis that began in the late 1960s prompted artists to begin creating pictures, paintings, and sculptures that represented the earth and this environmental ethic. In *Shifting Ground: Transformed Views of the American Landscape*, Rhonda L. Howard writes a commentary for an art exhibition held at the Henry Art Gallery from February 10–August 20, 2000, where she discusses the trend in earth art:

> The desire to return to the simplicities of nature and the concern surrounding environmental damage prompted multiple reactions from artists, reactions that continue into the twenty-first century. Working during the time when these environmental catastrophes and concerns first came to the fore were earth artists, also known as site artists, land artists, and environmental artists. Earth artists, concerned with the environment, embraced the planet itself as their medium. (2000, 43, 48)

Of course making new images is not the same as changing them, and even these environmental works of art present a certain view of nature. However, if perspective is an important part, then perhaps we can at least alter this one variable. But to alter perception, we must first get people to perceive. Although theories of ecosee need not be practiced, getting people to perceive, to pay attention to the billboards along the highway, is the challenge of a praxis of ecosee theory. While images may not work according to standard species of rhetoric (deliberative, forensic, epideictic), a study of environmental rhetoric allows us to see existing arguments into which these images become inserted.

Ecospeak was published at the brink of a new environmental movement. The early 1990s saw renewed anti-environmentalism among corporations, which began to apply their public relation resources to environmental issues. With the entrance of public relations firms, whose very business is rhetoric, Killingsworth's and Palmer's analysis of eco-speak could not have appeared at a better time. We now find ourselves in another backlash, where visual media play an ever-increasing role as we move from a literate to an electrate culture. As mentioned earlier, social networking sites such as Myspace and Facebook, as well as media-based sites that facilitate networking, such as YouTube, are providing online spaces where modern-day monkey wrench gangs can gather and discuss environmental issues, and do so through an image-based medium, something that simple e-mail and a listserv do not facilitate as well. Moreover, films such as Al Gore's *An Inconvenient Truth* form communities of viewers who discuss the film and environmental politics both in person and online. While electracy does not exclude literacy, we must develop a logic of how we see eco, not just how we speak it. The praxis of ecosee exists to the extent that it allows people to make practical decisions in their day-to-day lives, to see and feel obtuse meanings within any image, to understand how that image relates to all other images, and to use the rhetorical strategies needed to produce and argue with such images.

However, ecosee also should consider practical decisions that not only affect the viewer, but the impact that the viewer will then have on her or his relationships with the world. When an image portrays a white shark as a "monstrous, eating machine," it does so irresponsibly because this portrayal fails to take into account (or purposefully takes into account) the fact that people might fish out and destroy sharks at large. In his chapter "The Animals: Territory and Metamorphoses," in *Simulacra and Simulation*, Jean Baudrillard discusses how science attempts to make animals say they are not animals:

> Animals must be made to say that they are not animals, that bestiality, savagery—with what these terms imply of unintelligibility, radical strangeness to reason—do not exist, but on the contrary the most bestial behaviors, in the most singular, the most abnormal are resolved in science, in physiological mechanisms, in cerebral connections, etc. Bestiality, and its principle of uncertainty, must be killed in animals. (1994, 129)

Images of animals also attempt to make them say they are not animals, or that they are more than animal. When this bestiality cannot be erased, or the uncertainty cannot be fixed with text, then the image makes them into the econ of the white shark, which influences people to destroy them. The shark is no longer an animal but the hyperreal equivalent of death, no less than an image of the grim reaper. In our

Western metaphysics, this role makes sense for the shark, and it becomes anthropomorphized into a demon. As Orson Welles narrates for the 1975 *Jaws* trailer, "It is as if God created the devil and gave him . . . jaws" ("Trailers for Jaws," 2007).

I began by exploring where ecosee might borrow from Killingsworth's and Palmer's *Ecospeak* in an attempt to establish some general rules for environmental discourse and rhetoric. In their epilogue, these authors discuss the practical implications for their work and the necessity to develop new kinds of discourses, new audiences, new authors, and new conversations in order to break the discourse of ecospeak and actually create change. To do so, new rhetorics will need to transgress "boundaries of discourse communities" and develop "new personae," including these "new kinds of authors and audiences (real and ideal)" (1992, 279). As they posit:

> The social-epistemological and rhetorical adjustment we face is in every way comparable to the cognitive adjustment required to appreciate Stephen Schneider's joke: "Nowadays everybody is doing something about the weather, but nobody is talking about it." (1992, 280)

Although we still need to pay attention to ecospeak and focus on ways to change the rhetoric and conversations of environmental politics, we should be more concerned about those images and visual rhetorics that comprise ecosee; scholars are doing lots of work in the fields of visual rhetoric and new media, but few are talking about how these fields relate to the environment, fields that, as Orr suggests, are connected. And, on the other side, while many scientists are looking for solutions to environmental problems, few address the cultural capital stored within the environmental images used to communicate such problems. Mitchell discusses the important ethical role that a picture theory should play by asking the question: "What is our responsibility toward these representations?" (1994, 424). He responds:

> To begin with, we must see them as related to one another and to us. Although some of them may be "beyond our control," they are certainly not outside our field. In the case of the political correctness campaign, it is precisely our field that is in question. The new legitimations of racism and sexism are mediated by representations about which we have considerable expertise. And the representation of war and mass destruction in narratives that simultaneously erase the memory of Vietnam

and replace it with a fantasy replay of World War II should activate our responsibilities as preservers of the historical record and of cultural memory. In short, though we probably cannot change the world, we can continue to describe it critically and interpret it accurately. In a time of global misrepresentation, disinformation, and systematic mendacity, that may be the moral equivalent of intervention. (1994, 424–25)

Mitchell's field, as well as mine, is in an English department. However, if, as I argue, all images are environmental images, then all of us have a stake in understanding what images say, what we say with them, and how to make them say something more. The environment is not just interpreted by analyzing such images but is made, and participating in ecosee is literally "world making." Although theories of ecosee probably cannot change the ontological conception of "nature," it can at least help change the way we see its ontological surfaces, even if those surfaces are not acknowledged as such. While I do not necessarily agree that we can interpret nature "accurately," we can at least point out how ecosee might misrepresent, disinform, and systematically construct the environment to others who just assume that camels and nature exist. This "disinformation" occurs mostly at the obtuse meaning of pictures, the unsayable that plays on our emotions, with ekphrasis providing the chief means of extracting that meaning (even if it is a meaning that cannot be extracted). Since for the time being we must interpret such images through words, we should try to be critical and ethical about what we are really trying to say with these images and their descriptions. Lyotard points out, "The canonical phrase of Plato's poetics would be in sum: I deceive you the least possible" (1988, 22). A poetics of ecosee should demand nothing less.

Notes

1. When using the term *image*, I am referring to a broad array of visual stimuli. While I include such obvious examples as photographs and paintings under this definition, I also include moving images such as film and television, digital images found on the Internet, three-dimensional images such as sculptures and architecture, and presentational "images" such as habitat recreations in zoos and aquaria, but also natural parks and anything that is designated as a thing to be seen. My purpose here is to avoid taking the image as the representation of some object, and to consider that objects themselves often can be viewed as images.

2. For a more comprehensive examination of the relationship between discourse, nature, and the environment, see Dobrin and Weisser, 2002.

3. *Electracy* is a neologism developed by Gregory L. Ulmer to describe the kind of skill set necessary to make use of electronic media found primarily in the digital Internet (though not exclusively). In *Internet Invention*, Ulmer states that electracy "is to digital media what literacy is to print" (2003, xii). That is, while literacy provides the skills necessary for the technology of alphabetic writing, electracy (will) provide the skills necessary to make full use of the image, which, since the invention of photography, is now ubiquitous via the Internet.

4. Ulmer explains what electracy has to offer compared to literacy:

> What literacy is to the analytical mind, electracy is to the affective body: a prosthesis that enhances and augments a natural or organic human potential. Alphabetic writing is an artificial memory that supports long complex chains of reasoning impossible to sustain within the organic mind. Digital imaging similarly supports extensive complexes of mood atmospheres beyond organic capacity. Electrate logic proposes to design these atmospheres into affective group intelligence. Literacy and electracy in collaboration produce a civilizational left-brain right-brain integration. If literacy focused on universally valid methodologies of knowledge (sciences), electracy focuses on the individual state of mind within which knowing takes place (arts). (2007)

5. When examining ecospeak, Killingsworth and Palmer were investigating a rhetorical entity that already existed in the media and public discourse. Similarly with ecosee, I am suggesting that certain practices are already in use, but also that various institutions, including education, can create new practices.

6. For Barthes, the *obtuse meaning* (or *third meaning*) of an image exists at a level outside of language use. While other meanings of an image can be linguistically explained, such as the denotative meaning (what it is) or the connotative (what it is meant to represent/symbolize), the third meaning plays on emotion and is linguistically inexpressible. Barthes writes:

> The obtuse meaning is a signifier without a signified, hence the difficulty in naming it. My reading remains suspended between the image and its description, between definition and approximation. If the obtuse meaning cannot be described, that is because, in contrast to the obvious meaning, it does not copy anything—how do you describe something that does not represent anything? The pictorial "rendering" of words here is impossible. (1977, 61)

Electracy, potentially, will help make this inexpressible feeling associated with the image (sometimes referred to as the *punctum*) expressible by using new practices not available through alphabetic technology.

7. Derived from the term *icon*. Although other terms might be used for what I describe, I have developed these tentative terms for discussing elements of ecosee. Whether these terms prove useful for others is beyond their purpose

here. I use them in order to differentiate different aspects of the depiction of nature in images.

　　8. Also consider the ABEC Web site (http://www.balancedenergy.org/), which uses visual rhetoric to connote a sense of environmental friendliness with its multiple shades of green.

　　9. Raid® not only allows the home owner to kill bugs but to "kill bugs dead."

　　10. Barthes explains that when an image and a text are juxtaposed, the text "constitutes a parasitic message designed to connote the image, to 'quicken' it with one or more second-order signifieds . . . the image no longer illustrates the words; it is now the words which, structurally, are parasitic on the image" (1977, 25).

　　11. Eventually, through the skill set of electracy, we may be able to understand and express nature through other nonlinguistic aspects, such as Barthes's obtuse meaning.

　　12. This affinity also might exist as an identity if nature is thought of as itself an image. Park developers arrange and present the image of nature in deliberate ways. Zoos contain exhibits and "habitats" designed to give the viewer the expectation of an image of an animal's native environment. This expectation creates a hyperreal image of nature, one that negates any real difference between an image of nature as photograph and the "real" on site experience (read viewing) of nature.

　　13. In particular, I am thinking of N. Katherine Hayles's article, "Searching for Common Ground," in which she claims that a deconstructionist paradigm of nature would destroy environmentalism and asks the question, "If nature is only a social and discursive construction, why fight hard to preserve it?" (1995, 47).

Works Cited

Abbey, Edward. 1999. *The Monkey Wrench Gang*. Salt Lake City, Utah: Dream Garden Press.

Americans for Balanced Energy Choices. 2005. Advertisement. http://www. balancedenergy.org/abec/index.cfm?cid=7577, accessed March 5.

Baker, Steve. 2001. *Picturing the Beast: Animals, Identity, and Representation*. Urbana: University of Illinois Press.

Barthes, Roland. 1977. *Image, Music, Text*. Trans. Stephen Heath. New York: Hill and Wang.

Baudrillard, Jean. 1994. *Simulacra and Simulation*. Trans. Sheila Faria Glaser. Ann Arbor: University of Michigan Press.

DeLuca, Kevin Michael. 1999. *Image Politics: The New Rhetoric of Environmental Activism*. New York: Guilford Press.

Dobrin, Sidney I., and Christian R. Weisser. 2002. *Natural Discourse: Toward Ecocomposition*. Albany: State University of New York Press.

Foucault, Michel, 1973. *The Order of Things: An Archaeology of the Human Sciences.* New York: Vintage.

Gombrich, E. H. 1999. *The Uses of Images: Studies in the Social Function of Art and Visual Communication.* London, UK: Phaidon.

Hajer, Maarten A. 1995. *The Politics of Environmental Discourse: Ecological Modernization and the Policy Process.* Oxford: Oxford University Press.

Hayles, N. Katherine. 1995. "Searching for Common Ground." In *Reinventing Nature?: Responses to Postmodern Deconstruction,* ed. Michael E. Soulé and Gary Lease, 47–64. Washington, DC: Island Press.

Howard, Rhonda L. 2000. *Shifting Ground: Transformed Views of the American Landscape.* Vancouver, Canada: Hemlock Printers.

An Inconvenient Truth. 2006. Dir. Davis Guggenheim. Perf. Al Gore. Paramount.

Kennedy, Robert F. Jr. 2004. *Crimes against Nature: How George W. Bush and His Corporate Pals Are Plundering the Country and Hijacking Our Democracy.* New York: Harper Collins.

Killingsworth, M. Jimmie, and Jacqueline S. Palmer. 1992. *Ecospeak: Rhetoric and Environmental Politics in America.* Carbondale: Southern Illinois University Press.

Laclau, Ernest, and Chantal Mouffe. 1985. *Hegemony and Socialist Strategy: Towards a Radical Democratic Politics.* Trans. Winston Moore and Paul Cammack. London: Verso.

Latour, Bruno. 1993. *We Have Never Been Modern.* Cambridge, MA: Harvard University Press.

Lyotard, Jean-François. 1988. *The Differend: Phrases in Dispute.* Trans. Georges Van Den Abbeele. Minneapolis: University of Minnesota Press.

Mitchell, W. J. T. 1994. *Picture Theory: Essays on Verbal and Visual Representation.* Chicago: University of Chicago Press.

Nan, Wang Jia, and Cai Xiaoli. 2003. *Oriental Painting Course: A Structured, Practical Guide to the Painting Skills and Techniques of China and the Far East.* New York: Watson-Guptill.

Orr, David W. 1994. *Earth in Mind: On Education, Environment, and the Human Prospect.* Washington, DC: Island Press.

Plato. 1989. "Gorgias." In *Plato: The Collected Dialogues,* ed. E. Hamilton and H. Cairns, 229–307. Trans. W. D. Woodhead. Princeton, NJ: Princeton University Press.

Saint-Martin, Fernande. 1990. *Semiotics of Visual Language.* Bloomington: University of Indiana Press.

Sánchez, Raúl. 2005. *The Function of Theory in Composition Studies.* Albany: State University of New York Press.

Sonesson, Göran. 1989. *Pictorial Concepts: Inquiries into the Semiotic Heritage and Its Relevance to the Analysis of the Visual World.* Malmö, Sweeden: Lund University Press.

———. 19 Jan. 2004. "Visual Semiotics." *The Internet Semiotics Encyclopedia.* Lund University Press. http://www.arthist.lu.se/kultsem/encyclo/visual_semiotics.html, accessed 5 Mar. 2005.

Tomaselli, Keyan G. 1996. *Appropriating Images: The Semiotics of Visual Representation*. Højbjerg, Denmark: Intervention Press.

"Trailers for Jaws." 2007. *The International Movie Database*. http://www.imdb.com/title/tt0073195/trailers, accessed 2007.

Ulmer, Gregory L. 1994. *Heuretics: The Logic of Invention*. Baltimore, MD: Johns Hopkins University Press.

———. 2003. *Internet Invention: From Literacy to Electracy*. New York: Longman.

———. 2007. "Electracy and Pedagogy." *Internet Invention: From Literacy to Electracy*. University of Florida. http://www.web.nwe.ufl.edu/~gulmer/longman/pedagogy/electracy.html, accessed 8 June 2007.

White, Daniel R. 1988. *Postmodern Ecology: Communication, Evolution, and Play*. Albany: State University of New York Press.

Williams, Joy. 2001. "Save the Whales, Screw the Shrimp." In *Ill Nature: Rants and Reflections on Humanity and Other Animals*, New York: Lyons Press.

2

Ecoporn

On the Limits of Visualizing the Nonhuman

Bart H. Welling

> Nature photographs have become something of a problem.
> —Joy Williams, "Save the Whales, Screw the Shrimp"

It is no secret that mainstream visual representations of unbuilt landscapes and nonhuman animals can bear more than a passing resemblance to pornographic images of humans, especially women. José Knighton and Lydia Millet, among others, have helped popularize a name for this phenomenon: ecoporn. Without addressing the concept of "ecoporn" per se, figures as diverse as Carol J. Adams, Terry Tempest Williams, Susan Griffin, Susanne Kappeler, and Charles Bergman have shed light on the issue of how pornography and visual representations of nature (photographs and documentary films in particular) can code the viewer's eye not just in similar but, as I will argue here, in deeply interrelated ways: as solitary, central but remote, omniscient, all-powerful, potentially violent, pleasure-taking, commodifying, an all-seeing but simultaneously invisible consuming male *subject* to its marginalized, decontextualized, powerless, speechless, unknowing, endangered, pleasure-giving, commodified, consumable female *object*. The dangers of the "god-trick" (to adapt Donna Haraway's felicitous phrase [1991, 191]) are just as readily apparent in environmental visual regimes as in pornography. The possibility of eros—whether defined as intersubjectivity, intercorporeality, multisensory dialogue, or something else—can collapse in the face of the tyranny of the visual.

Despite these real correspondences and potentially damaging consequences, ecoporn (as trope, as mode of representation, and as ethical problem) has received surprisingly little sustained theoretical attention. In this chapter I attempt to build on the work of the thinkers mentioned earlier by exploring the dialectical relationship between ecopornographic representations and human attitudes, beliefs, and behaviors vis-à-vis the nonhuman world. I anchor my observations in an analysis of ecoporno-graphic representations of nonhuman subjects like the Florida panther before concluding with some of the less ecopornographic possibilities embodied in such postmodern approaches to the problems of represent-ing the nonhuman as the work of the British duo Olly and Suzi. At the same time, I want to help define the limits of ecoporn as a concept. Are we dealing with a metaphor, a metonym, or exactly the same aesthetics, logics, and narratives in ecoporn as in human porn? How large a role does human sexual desire play in ecopornographic visual discourses? Are airbrushed shots of the Tetons qualitatively different from filmed scenes of battling grizzlies? How do representations of animal sexuality fit in? I can only begin to answer these questions here; the main thing for now is to raise them. They are vital because they have a bearing not just on how we see nature but on how we make homes for ourselves in a world full of other life-forms.[1]

First, some definitions are in order. The former advertising executive Jerry Mander, a Deep Ecologist, was one of the first figures to employ the term *ecopornography*, doing so in a 1972 article by that name. For Mander, "ecopornography" is basically synonymous with the more popular term *greenwash*, defined by the Oxford English Dictionary as "Disinformation disseminated by an organization, etc., so as to present an environmentally responsible public image [. . .]." By this definition the difference between ecopornography and more ecologically sound uses of images of nature is crystal clear. A Westinghouse advertisement showing boaters and swimmers frolicking within a few hundred feet of a "clean, safe" nuclear power plant is inherently ecopornographic (see Mander, 1972, 51), while an ad sponsored by the Friends of the Earth, featuring an illustration of a live mink and the signatures of prominent individuals who have renounced wearing fur (see "Exhibit" 1972, 58), is not. In Mander's view, the environmental credibility of the organiza-tion propagating a given image deserves more scrutiny than the rhetoric of the image itself. Mander briefly flirts with the idea of anti-ecopor-nographic alliances between environmentalists, minority activists, and feminists (1972, 55), but he does not explore the deeper congruencies I observe between human pornography and ecopornography. Nor does he address the problem of structural continuities between Shell Oil's pictorial vision of nature, for instance, and the Sierra Club's.[2] Ecopor-nography is like "real" pornography, Mander implies, because it masks sordid agendas with illusions of beauty and perfection. In his hands, "ecopornography" stays safely within the realm of a catchy but theoretically unfruitful metaphor.

In recent years José Knighton and Lydia Millet have offered more nuanced views of "ecoporn," focusing less on the distinction between corporate and environmentalist motives than on the damaging effects that idealized visualizations of nature of *any* provenance can have on viewers' attitudes toward the ecosystems upon which these images are based. In fact, both authors draw special attention to the counterproductivity of images which, instead of educating the public about ecosystemic abuses and mobilizing dissent, anesthetize viewers by constructing an illusion of a pure, safe (at least for now!), friendly, ahistorical, depopulated, monolithic Nature located in an indeterminate space far from the viewer's degraded backyard but nonetheless, paradoxically, subject to the viewing eye's private, voyeuristic control. Significantly, while both Knighton and Millet do indulge in some purely metaphorical wordplay—"To add insult to injury," Millet writes, "[mainstream environmentalist images of nature] don't even get us off" (2004, 147)—they also help lay the groundwork for a definition of ecopornography centering on metonyms more than on metaphors, on the homologous rather than merely analogous "kinship" (Knighton 2002, 170) between ecopornographic and standard pornographic technologies, iconographies, and ways of disciplining viewer responses.

Over the next few pages I hope to contribute to an expanded theory of ecopornography by placing observations such as Knighton's and Millet's in dialogue with feminist critiques not just of traditional pornography but of the patriarchal logic undergirding the linked exploitation of women and nature. Ecopornography, I contend, is part and parcel of the pornographic continuum. While American environmentalism, true to its roots in the larger culture, sometimes calls on us to imagine a "clean" Nature as the antithesis of "dirty" human sexuality and a diseased human nature in general, anatomizing ecoporn *as* porn can help us interrogate the androcentric, ethnocentric, and anthropocentric assumptions that—despite appearances to the contrary—animate ecoporn, perpetuating Nature/Culture dichotomies that sabotage environmentalism's attempts to inspire wider publics. This is not to say, of course, that only environmentalists produce ecopornography. Jerry Mander's definition should be expanded, not rejected. It *is* always worth asking whether a given ecopornographic representation is being used by an organization trying to greenwash its image or, rather, by one hoping to promote environmentally friendly policies or at least appreciation for the natural world. But over thirty years after the first Earth Day, the more pressing issue now is that, without reading the text accompanying an image or listening to the voice-over of a documentary, *you often cannot tell* whether that family of otters is "smiling" because Exxon decided to spill a little less oil on them in the future or because Defenders of Wildlife stepped in to clean up Exxon's latest mess. Either way, if you are a typical media-saturated resident of the United States, your response will likely be the same: you feel mildly pleased with the world, flip the page, change the channel . . . or fall asleep. This apathetic consumerist response should be exactly what environmentalists work to unsettle, not promote. The question of how environmentalists can (re)visualize nature, how they can use visual rhetoric to help transform passive consumers of resources into thoughtful inhabitants of ecosystems, has never been more urgent.

Before going any further, though, let me address some potential points of confusion as directly as possible. Since by my definition eco-pornography concentrates on a certain type of image in which humans are for the most part conspicuously absent, I will not be dealing with representations of "zoosexuality," or with pornography featuring human sexual contact with plants or inorganic matter.[3] My concern here is with a type of pornography that is more subtle than standard human porn but just as widespread—and in some ways just as pernicious. I will, how-ever, touch upon the processes of pornographic substitution by which landforms and nonhuman animals come to stand in for women in the public eye, and, on the other hand, the processes by which women are naturalized and animalized. Thinking about ecopornography and standard heterosexual pornography as related categories on the same spectrum rather than totally independent phenomena can illuminate the chiastic nature of their relationship. Likewise, this holistic approach can help us understand the function of desire and power in ecopornography as dynamic translations rather than abstract simulacra, of the interworkings of desire and power in more traditional pornographic contexts. In other words, instead of bowing to the obvious objection that ecopornography does not literally "get us off," and thus cannot properly be considered pornography, I would suggest that ecoporn-*as*-porn places the viewer in the same asymmetrical, sexualized relationship to its subjects as standard pornography, even if its primary goal is not sexual arousal. Needless to say, the question of whether it is arousal—and not, for instance, patri-archal control of women's sexuality and/or the making of vast sums of money—that constitutes the "primary goal" of pornographic images of *women* is itself a highly vexed one. Pornographic desire can train its gaze on more than just human bodies, and can train the viewer's gaze in ways too subtle to be measured with a plethysmograph.

What, then, does ecoporn look like? Where is it found, and by whom is it produced and used? To paraphrase U.S. Supreme Court Justice Potter Stewart's elliptical 1964 observation on pornography, how would we know ecoporn if we were to see it? At this point I cannot resist the temptation to rush in where Justice Potter feared to tread, offering an extended working definition of ecoporn in the hope that it will be debated and improved on by others.

Ecopornography is a type of contemporary visual discourse made up of highly idealized, anthropomorphized views of landscapes and nonhuman animals. While these images often are composed or manipulated to stress their subjects' innate similarities to the human body and to human social and power structures (such as the nuclear family, patriarchy, and the nation-state), the images work to conceal both the material circumstances of their creation by humans and whatever impact humans may have had on the landforms and animals they depict. This hidden impact includes the stress experienced by animals (not all, but some) that have been chased, shot with tranquilizer darts, domesticated, spray painted, "posed," or otherwise disturbed to obtain ecopornographic images.[4] Similarly, ecoporn hides the costs to ecosystems of being visited and popularized by those who photograph and film them (although some threatened ecosystems undoubtedly benefit from any outside attention, ecopornographic or otherwise, and differences in *how* places are visited—cruise ships or backpacking tours?—must obviously be factored in). Ecoporn also conceals the doubly invisible forms of damage inflicted on the *non*represented, nonphotogenic landscapes that are logged, mined, dammed, polluted, or otherwise exploited to provide the materials and energy required for producing and distributing images of more visually appealing places.[5] Instead of exploring such problems of representation, ecoporn supplies viewers with a fantasy of benign but total visual power over these nonhuman creatures and habitats that are both comfortingly humanized and pleasingly "untainted" by humans. ("Visual power" often translates in ecoporn as a sense of private ownership: the illusion, for instance, that one is the only human gazing at, visually possessing, the gorgeous sunset-washed ocean off of Waikiki, where the beach behind the photographer is most likely very crowded.)

Ecoporn is designed for quick, easy, visual consumption. It is pornographic and not simply *like* porn, for at least three interrelated reasons.

1. It traffics in pictorial versions of the same land-as-woman tropes that, as Annette Kolodny, Anne McClintock, and others have shown, have done much to authorize the genocidal oppression of native peoples and the colonization of their lands by European settlers and the nonhuman

animals, plants, and diseases they have brought with them. In other words, ecoporn perpetuates ways of seeing feminized Others that have been instrumental in facilitating countless acts of violent expropriation. Gregg Mitman's account of the role of wildlife films in constructing the image of Africa as an "untouched animal paradise" (1999, 197), and thus in paving the way for the displacement of indigenous groups such as the Maasai by white conservation organizations and ecotourists, offers powerful support for the idea that this process is far from exhausting itself. The effects of ecoporn are unquestionably less catastrophic than those of earlier iterations of the land-as-woman idea, but we are dealing with differences in degree rather than in kind; the deep structures of representation are the same. And, crucially, these mirror the structures by which traditional pornography equates the female body and women's sexuality with exploitable or domesticable nature (kitten, beaver, pussycat, bitch [see Adams 1990, 61]) and everything male with non-natural, nature-transforming culture.

2. To borrow a key phrase from Susanne Kappeler's (1986) *The Pornography of Representation*, ecoporn places the viewer in the role of the "male surveyor," the all-seeing male subject to Nature's unseeing, aestheticized female object. As Kappeler demonstrates in rewriting John Berger's essay "Why Look at Animals?" as "Why Look at Women?" (her sixth chapter-length Problem), representations of women and of nonhuman animals in modern Western culture are interchangeable in truly "uncanny" ways (63), and the effects of being objectified visually, whether in zoos or in peep shows, can be sickeningly close for women and animals. It would be entirely consistent with Kappeler's argument to apply her observation, "The fundamental problem at the root of men's behaviour in the world . . . is the way men see women, is *Seeing*" (61, emphasis added), to the role of ecopornographic vision in denying the capacity for active, independent vision—and, by extension, other kinds of agency—to nonhuman life-forms. (But do all forms of pornography *necessarily* objectify the complex beings they represent? Could not ecoporn be defended with the argument that it, along with standard pornography, has the capacity to empower as well as to degrade? While a version of the first question has been debated in and out of feminist circles for many years, the second question clearly merits further study; for now, I will have to maintain that the costs of ecoporn frequently outweigh the benefits.)

3. Finally, ecoporn is pornographic because, to a disturbing degree, it has been moving in recent years beyond "[t]arted up" images of flesh-colored landscapes (Millet 2004, 147) and big cats captured in the seductive poses of *Playboy* models (themselves sometimes portrayed in

images set up and manipulated to fuel the fantasy of what Kappeler calls the "wild animal-woman, the sexual beast" [1986, 75]) to a truly hard-core obsession with explicit sexuality and violent death, microscopically examined. That the new "animal snuff films" (Mitman 1999, 208) are branded "educational" and shown openly during prime time by various television networks does not render them less *pornographic* than films featuring voyeuristic representations of sexual violence against women—even if most viewers would unhesitatingly rate them less obscene.

Judging by the complexities of these criteria, "ecopornography," like "pornography" itself, may seem like a hopelessly baggy concept. What on earth do the World Wildlife Fund's photographs of adorable and "adoptable" baby pandas, for example, have to do with the Discovery Channel's *Anatomy of a Sharkbite*? Actually, the history of Western ideas of, and interactions with, nature is replete with examples not just of how our approaches to nature are governed by various dualisms—Man-as-Son/Man-as-Rapist, Nature-as-Dystopia/Nature-as-Utopia (Howling Wilderness/Eden), civilization/savagery, victim/victimizer, destroyer/savior—but of how quickly and dramatically the terms constituting these binarisms can switch places. This volatility is particularly evident in the history of endangered species. Of wolves in the United States, for instance, Andrew C. Isenberg writes, "Feared and reviled as loathsome and cowardly killers at the outset of the [twentieth] century, they have come, by the century's end, to symbolize the possibility for holism and integrity not only in the American environment, but in American culture" (2002, 49). As Nigel Rothfels notes, sharks are undergoing an image restoration of their own, as the Discovery Channel tries to change the focus of *Shark Week*, and Peter Benchley (author both of *Jaws* [1974] and of the more complex meditation *Shark Trouble* [2002]) teams up with the National Geographic Society in print and on film to transform great white sharks in the popular imagination from killing machines to paragons of evolutionary fitness. Such metamorphoses should be viewed in the context of measurable historical trends, such as the drop in numbers of a given species due to overpredation and habitat destruction by humans. But I would add that it is equally important to pay attention to the instability of the psychic structures that help drive, and do not merely respond to, these tangible changes. "[I]t was okay to demonize an animal [in the 1970s], especially a shark," Benchley argues, "because man had done so since the beginning of time, and, besides, sharks appeared to be infinite in number" (qtd. in Rothfels 2002, viii). The balance tips, and the infinite becomes the vanishing; with predators, the hunter becomes the hunted, and what formerly endangered us is now endangered *by* us. Likewise, former wastelands, such as deserts, are enshrined as places of sacred retreat from the all-devouring spread of human chaos. What is not "okay" about this seemingly positive revolution in how we visualize and live with the nonhuman is the degree to which it is *not* a revolution but rather, as Susanne Kappeler might say, a *coup d'etat*.[6] The Nature/Culture scales tip too easily in either direction, not just one; we are content to see nature (temporarily) occupy the superior position without interrogating the deeper imbalances and schisms at the heart of how we represent, see, and live in the world. Rothfels detects a troubling oscillation in National Geographic's rhetoric (both linguistic and visual) between the "old" and "new" images of the great white shark, an alternation that comes to feel almost like a nervous breakdown in sentences like this: "With cinematography that takes you literally into the gaping mouth of this fearsome killer, it's a surprising new look at one of the largest, most fascinating predators ever to swim the seas—or haunt our imaginations" (quoted in Rothfels 2002, x). Instead of using the anniversary of *Jaws* to pose hard questions about how our dualistic ways of representing the nonhuman translate to violently contradictory behaviors, the Benchley/National Geographic projects fall back on what Rothfels characterizes as the "basic conceit of most nature films": that "no one (much less an extensive crew) stands behind the camera, and that what we see before the camera is an unmediated, unedited experience of 'Nature' " (ibid.). Not surprisingly, Rothfels adds, Benchley and his colleagues participate in some contradictory behavior of their own as they entice great whites to leap out of the water in pursuit of seal decoys without asking whether this might rob them of energy that should be used in pursuing real pinnipeds (ibid.). It would seem that ecoporn blinds its producers as well as its consumers to the costs of visualizing Nature as what, and where, Man is not.

It must be acknowledged, of course, that some ecopornographic representations of nature *do* include humans. I am fully aware that many wildlife documentaries, for example, devote less attention to the ostensible subjects of the films than they lavish on the filmmaker's struggles to master new technologies, travel to remote locations, overcome disease and self-doubt, gather truthful information from the natives (who, we may be informed, are suspected or "converted" poachers), and prepare himself and his equipment for difficult and unprecedented feats of camera work ("For the first time ever . . . !"). At least Benchley and others admit that they used decoys to obtain the shots of breaching sharks; does not their candor offer a glimpse into the "material circumstances of [the images'] creation by humans," satisfying what I see as the public's need for greater self-reflection on the part of image makers? Barbara Crowther would likely agree with me that it does not. In her critique of natural history programs from a feminist vantage point, Crowther identifies the "how-they-shot-it" format as a recent offshoot of three narratives that have dominated television wildlife shows practically from the beginning: the masculinist quest narrative, the birth-to-parenthood (male) life-cycle story, and the pervasive narrative about how science triumphs over (female) nature's mysteries (see Crowther 1995, 143). What should be added to Crowther's account is that, as I have tried to suggest in my brief cartoon sketch of the genre, these male heroic narratives in nature programs often follow hard on the heels of hunting plots and colonial discovery narratives. (Missionary narratives also are lurking in the not-too-distant background, as the emissaries of ecological enlightenment are shown persuading locals to abandon illegal hunting in favor of ecotourism). The only real difference is in the end result: instead of claiming a new territory for the Empire or returning from the wild with the head of a greater kudu, the hero must obtain shots that leave "untouched" what was targeted, and not just any shots but fantastic images of places, creatures, and phenomena that "have never been seen before." If the hero appears in these scenes, à la the late Steve Irwin (*The Crocodile Hunter*), an Australian legatee of the American Marlin Perkins (*Wild Kingdom*), it is generally to drag an uncooperative snake out of a hole, dodge a charging rhino, or otherwise demonstrate his unparalleled adroitness in managing wild creatures that should never, ever be approached by ordinary viewers. (Filling in for these incompetent consumers in Irwin's program is his wife, Terri, who usually is shown tripping over ropes, scrambling to find a bag in which "the Hunter" can stuff his latest catch in order to "save" it from humans, or smiling in the background while her husband strokes a given reptile, crooning "Isn't she gorgeous?" partly to the camera and partly to the struggling animal).[7] If, as often happens, the filmmaker/hero does *not* appear in the ecopornographic shot he has worked so hard to get, then we are supposed to imagine that we are witnessing virgin wilderness, unmanipulated wildlife, a world apart from our degrading touch. However, whether the hero is the solitary *adventurer* confronting perils where other people, particularly women and native pastoralists/"poachers," are shown to be out of place, or the *discoverer* relaying his images of natural perfection to a civilization that has fallen from such grace, the emphasis is clearly on phallocentric closure (Crowther 1995, 132), on a distant female Nature that can only be penetrated, literally and figuratively, by male figures, male technoculture, and male narratives. The fetishization of our technologies of seeing leads not to critical introspection about the limits of human vision but to something like its opposite: an illusion both of solitary, unmediated experience and of complete visual power.

By the same token, the sense of ambivalence that one would expect to arise from the filmmakers' entanglement in these contradictions (domesticating what they want to keep wild, "raping" what they claim to experience as "virgin") has been diffused into an ever-more-frantic search for new frontiers, new Edens—internal (ice caves, undersea volcanoes, reproductive tracts) as well as external. In his essay "Nature's Voyeurs," environmental writer Richard Mabey compares contemporary wildlife filmmakers to Enlightenment "knowledge vandals" such as William Derham, who, in 1711, issued this admonition to readers: "Let us ransack all the globe, let us with the greatest accuracy inspect every part thereof, search out the innermost secrets of any of the creatures [. . .] pry into them with all our microscopes and most exquisite instruments, till we find them to bear testimony to their infinite workman" (qtd. in Mabey 2003). Mabey does not comment on the gender politics of this statement, but—in its potential grammatical confusion of "creatures" with "instruments," and thus of Man with the "infinite workman," God—it demonstrates almost too clearly how, in what Haraway calls an "ideology of direct, devouring, generative, and unrestricted vision, whose technological mediations are simultaneously celebrated and presented as utterly transparent" (1991, 189), the viewer is seduced into believing that his apparently "infinite" powers of sight translate to an equally godlike *knowledge* of the feminized objects of his gaze. This seemingly perfect understanding in turn facilitates new god projects as we literally revisualize (transform according to the dictates of human vision) the landscapes and species that do not conform to our ecopornographic ideals. Africa must be made to look like *Born Free*—and, for good measure, central Florida should be revisualized as *The Lion King*, complete with a fourteen-story concrete Tree of Life. If pushed too far, Derham's vision collapses into a solipsistic nightmare in which the more desperately we try to discover Nature, the more we see our "most exquisite instruments" staring blankly back at us.

Not that we can always recognize this trap when we fall into it. Ecopornographers have become singularly adept at crafting images of animals and places that *appear* to be reciprocating our affections, gazing back at us longingly (baby seals, big cats), or inviting us to lose ourselves in primal bodyscapes (rose-colored peaks, womblike slot canyons of the American Southwest) that visually recapitulate the pastoral vocabulary employed by early explorers such as Sir Walter Raleigh, who described Guiana in 1595 as "a countrey that hath yet her maydenhead, never sackt, turned, nor wrought" (qtd. in Kolodny 1975, 11). As more and more critics are showing, our ecopornographic fantasies of consensuality—such as the colonists' accounts of the "virgin" New World and, yes, the illusions of intimacy created by standard pornography—depend on conscious acts of erasure, exclusion, and distortion, not to mention complicity in this deception on the audience's part.

Consider the case of the Florida panther (*Puma concolor coryi*), the state animal of Florida and the charismatic "[m]ost endangered large mammal in North America" ("Florida Panther" 2005). It does not tend to show itself willingly to humans. Chickasaw writer Linda Hogan devotes several passages in her 1998 novel *Power* to the animal's wondrous ability to watch humans without itself being seen. Biologists go to extreme lengths to monitor Florida panthers, tracking them in the same way—and often with the help of the same people—as the panther's mountain lion relatives (*Puma concolor*), which are killed legally by trophy hunters in Arizona and other Western states. After being chased, treed, shot with tranquilizer darts, and hoisted down to the ground, Florida panthers often are discovered, in Charles Bergman's words, to be "sick, scrawny, and scarred" (2003, 99). Bergman also notes that the photographic image of their state mammal, with which all Floridians are familiar—of a noble, consummately wild predator at rest, gazing fearlessly into the camera lens or off into the distance—is generally obtained using tame panthers in carefully controlled settings (93). These complicated realities are not, of course, what Floridians want to see. Florida's love for its panther, fueled by ecopornographic illusions of mutuality, has made the "Protect the Panther" license plate the top-selling specialty license plate in Florida for the past few years, despite the irony involved in emblazoning a pro-panther message and the image of a healthy panther on hundreds of thousands of the vehicles that collectively pose one of the greatest threats to the panther's survival.[8] Funds from the sale of Florida's "Conserve Wildlife" license plates—which were originally sponsored by a public-private partnership, including Defenders of Wildlife and the Florida Chapter of the Sierra Club—support conservation, education, research, and law enforcement but also "watchable wildlife initiatives" and the creation of educational Web sites such as Florida Panther Net, where children can see images of Florida panthers relaxing in trees and striding majestically through Edenic woods with no freeways, subdivisions, biologists, or wildlife photographers in sight. While Florida Panther Net also shows photographs of panthers killed by hunters and cars, and there *is* something to be said for the role that images of charismatic megafauna can play in contributing to a "love of nature" in children, I am troubled by the circularity, willful naïveté, and obsession with visuality entailed in this kind of love when it persists into adulthood. Where do living, wild Florida panthers—not just "state symbols"—fit in? How do *they* see *us*? How can the complexities and indeterminacies of real human-panther encounters be represented more faithfully? How might our survival be linked with, and not shown as antithetical to, theirs? How must we change our behavior to coexist with them in mutually beneficial ways in the "mortal naturecultures" both humans and panthers actually inhabit?[9] The visual rhetoric of panther love, which has its roots in "the [Euro-American] vocabulary of a feminine landscape and the psychological patterns of regression and violation that it implies" (Kolodny 1975, 146), denies the need for these sorts of questions and may thus stand in the way of workable solutions.

Crucially, a nonvisual vision of the Florida panther such as Hogan's *Power* can do much to unmask ecoporn's failings; it also can unsettle the linked structures of oppression out of which ecopornography grows. After participating in the sacrificial killing of a Florida panther by her mentor, Aunt Ama, Hogan's narrator Omishto—a young member of the fictional Taiga tribe—finds herself transformed in the eyes of the white community from a model Indian into a "Swamp Nigger" (1998, 211). "I think of [the panthers]," Omishto observes during Aunt Ama's trial for killing the endangered animal,

> [. . .] the cats out there in the cypress and mangroves and swamps humans aren't meant to enter, not most humans anyway, though it seems to me like a natural place. And I think of the cats killed by cars. A dozen of them since the highway went in. The high school team is named after them. They are mascots, nothing more. No one wants them around, but they like to see them just the same. [. . .] There's no place human wants will let them be. (Hogan 1998, 123)

Omishto's words and experiences highlight the fatal contradictions inherent in the Euro-American approach not just to panthers but to the original human inhabitants of the "New World"; panthers and Native Americans alike become subject to gross discrimination when they insist on something more than being photogenic mascots, on more than *being seen*. Hogan's novel calls on readers to revise their notions of endangerment to include human cultures as well as nonhuman species. At the same time, it rejects the idea that endangered cultures or species are simply passive victims waiting to be saved or destroyed. Hogan revisualizes animals and ecosystems as full, active, and powerful participants in a transformative dialogue that depends on different ways of seeing than those inculcated in viewers by ecopornography. Readers learn that the "objective" view of the Florida panther offered by the state and even by environmentalists conceals a host of diverging human conceptions of the panther animated by differences in gender, religion, age, class, tribal versus federal law, and other categories. Moreover, the colonizing gaze that attempts to impose a monolithic, transparent meaning on the Florida panther—the victim rather than the victimizer, for now, in this unstable dichotomy—is shown from the beginning to be closely aligned with the mechanisms of patriarchal and sexual abuse. At the start of the book Omishto feels endangered not just by Euro-American culture but by her stepfather, Herm, whose eyes are "always looking too much in places they don't belong" (7). Omishto, whose name means "One Who Watches," gains in power as she learns to "stand up tall and look back at [Herm]" literally (208) and to "look back" figuratively at the inner workings of the religious, educational, and juridical power structures that attempt to destroy her sense of Native identity. Simultaneously, under the tutelage of Ama, who is able to "exchange glances" with the panther (16) and experience its full presence in dreams rather than merely see or be seen by it in person, Omishto comes to embrace an understanding of the animal's personhood that is not based on human gender constructions ("Sisa [the Florida panther] was the first person to enter this world. *It* came here long before us" [15, emphasis added]) and of a sacredness not founded on dualistic notions of body and soul, earth and heaven. It would be wrong to imply that the book does not pose serious interpretive problems for Native and non-Native readers alike, but in my view these unanswered questions count as great strengths, since they introduce readers to the necessary debates and complex histories that ecopornography stubbornly refuses to acknowledge, let alone engage in.

As should be clear from my definition of ecopornography, the usual images of solitary, resting panthers (and, one might add, jaguars, leopards, tigers, lions, and other big cats) are ecopornographic in spite of the fact that they are not *calculated* to arouse human sexual responses—that, indeed, they often are produced and distributed as safe, asexual images with an audience of children in mind. I can imagine many of my fellow parents objecting loudly to the idea that there is any connection whatsoever between *Playboy* and Florida Panther Net. The trouble with this kind of "intentionalist" approach, as with Jerry Mander's definition of ecopornography, is that it would distract attention from the rhetoric of the images themselves. And, as I see it, this rhetoric taps into at least two problematic traditions in Western culture at once. One is the association of animal skins and fur with wealth and female sexuality (see Fudge 2002, 52–65). Another is the long history of the female nude in visual culture. Ecopornographers are able to combine the traditional iconography of the female nude—with the unclothed but nonetheless coyly self-concealing subject lying within a short distance of the observer—with the fashion appeal of fur. Ecopornographic subjects are thus objectified not once but many times over. When idealized and anthropomorphized big cats, in particular, are shown at rest, divorced from the ecological contexts in which they live and either (apparently) unaware that they are being photographed or, by calmly looking at the lens, seeming to give their consent to be *represented* both in a technological sense and with animal rights or conservationist projects in mind, the viewer (despite the millennial rhetoric of the linguistic texts that tend to accompany these images) is most definitely not placed in the position of a fellow animal—working hard to catch a glimpse of a fast, powerful predator that under truly wild circumstances would not welcome the attention. Rather, we are cast in the role of voyeurs, potential destroyers, and/or potential saviors of the compliant or unknowing—*not* unwilling or independently "willed"—animal victim. And those who view themselves as rescuers are simultaneously positioned as consumers, albeit of a more rarified breed than those aligned with the forces of high fashion and habitat destruction. We consume images. The viewer's decision often is framed as a choice between consuming beautiful "shots" of endangered animals and actually (if vicariously) shooting them.[10] The "right choice"—whether it requires purchasing a license plate or sending a check to save a beautiful leopard from poachers—helps facilitate an endless and exponentially growing cycle of visual consumption involving wildlife calendars, T-shirts, posters, coffee mugs, toys, television networks, IMAX films, and countless other technologies of representation, along with zoos, museums, and ecotourism. In the end, despite the large sums ecoporn helps raise in the name of animals and the environment, its rhetorical focus is less on nonhuman subjects—which suffer from our rage to look at them in ways that ecopornography not only does not show but must keep hidden in order to survive—than it is on Man's pleasure, Man's power, Man's control.

That ecopornographic pleasure and power are in fact pornographic becomes even clearer in the context of the recent explosion of violence and sexuality in nature documentaries. Like shots of indigenous (i.e., more "animalistic") women's breasts as opposed to the wayward nipple of a performer such as Janet Jackson, whose humanity viewers have been trained to consider full and equal to their own, footage of nonhuman animals killing other animals or engaging in nonsimulated penetrative sex slips past parents, network censors, and the FCC because of the vast gulf that is assumed to exist between nature and culture, humans and animals, science (or what passes for science in most documentaries!) and its objects of study.[11] While the rapid growth of hard-core ecoporn is relatively new, the dichotomies underwriting it are not. Matt Cartmill relates that the "doggedly clean-minded" Walt Disney, one of the early pioneers of the wildlife film, enjoyed showing his office guests footage of mating animals despite thinking of standard Hollywood films as " 'putrid, pornographic shit' " (1993, 177). And yet *Bambi* itself (in Cartmill's reading [161–88]), along with the popularity of hard-core ecoporn in our day, belies the validity of these absolute divisions. Why would viewers find representations of "animal" sex and violence so intensely fascinating if these did not speak powerfully to our paradoxical need both to give expression to our animal selves and to control other animals and repress our own animality?[12] Like most documentary viewers, I would have to stress that representations of nonhuman sexuality and violence are not meant to be precisely *as* arousing as corresponding human representations, because the capacity for vicarious participation is understandably diminished.[13] But, again, more attention needs to be paid to distinctions in degree rather than kind between ecopornography and standard porn. If pornography is defined with a view primarily to the oppressive, androcentric, sexualized (not necessarily "orgasm-producing") power relationships it visualizes (sees, stages, perpetuates) in various media and in everyday life, then narratives about male animals that appear to do nothing but fight with each other, kill other animals, and mate in slow motion with docile females, whose sole purpose, in turn, is presented as that of breeding a new crop of males, are clearly—not "nearly," as is often argued—pornographic. Carol J. Adams and Josephine Donovan suggest that such narratives "work to reinscribe male-supremacist ideologies, both in promoting a view of nature as dominated by aggressive and violent males, and in sanctioning human male behavior that follows this model." "[This model] is designed," they continue, "to arouse fear in women and to promote their sense of needing men's protection" (1995, 6–7). Whether this design is conscious or not, more work is definitely needed on the reciprocities between ecoporn and male and female attitudes and behaviors toward Human Nature as well as Nature. What role might ecoporn play in the well-publicized sexism experienced by women both on the trail and in the boardrooms of leading environmentalist organizations? How does ecoporn (as exemplified not just in documentaries but, perhaps, in Disney's *Pocahontas* and *The Lion King*) affect boys' treatment of girls and girls' self-image? How do internalized ecopornographic gender norms help determine the kinds of wildlife films that get funded, and the ways in which these are made? These questions grow in importance not just as our environmental crises worsen but as the lines between ecoporn and standard pornography become more and more blurred.

For some of the most aggressive examples of this blurring, we can look in what may sound like a surprising place: the advertising campaigns of People for the Ethical Treatment of Animals (PETA). First, it is important to note that some of PETA's efforts, such as the

Animal Liberation Project, work in a potentially transformative way by exposing the deep connections between the abuse and exploitation of animals and the oppression of human Others by white males. The online exhibit for the Animal Liberation Project, for example, couples historical images of humans being oppressed with shockingly similar photographs documenting the abuse and exploitation of nonhuman animals.[14] The exhibit does not merely shock, however, as it provides viewers with an introduction, however abbreviated, to the arguments of Peter Singer. A nearly antithetical approach can be seen in the "We'd Rather Go Naked Than Wear Fur" campaign. In many of these ads, naked female celebrities strike soft-core pornographic poses to offer themselves for visual consumption—in exchange, it seems, for the safety of animals used for various fashion products (see "Famous Fur Foes" 2005). In one ad, *Baywatch* actor Gena Lee Nolin stretches out on a log, her unclothed body painted like a snake, beneath the caption "Exotic Skins Belong in the Jungle—Not in Your Closet." The "We'd Rather . . ." campaign is one of many headline-grabbing projects linking vegetarianism, fur-free fashion, and so on with sexual performance and sexiness.[15] The problem with these deliberate hybridizations of traditional animal rights advocacy and pornography, in my view, is that they reify precisely what animal rights and environmentalist groups should be most interested in subverting: the ways of seeing that have authorized all kinds of oppression by animalizing/naturalizing the human Other and by feminizing the nonhuman. The PETA/Nolin ad recalls countless statements not just by pornographers but (as Susan Griffin might note) by male authors of Scripture and theologians about the snakelike nature of women and the treacherously feminine nature of undomesticated Nature, that is, "the Jungle," even as it attempts to substitute one detachable commodity, the photographed "exotic skin" of a woman, for another, the skin of a reptile.[16] This skin trade self-destructs because it tries to fight speciesism with sexism.[17] PETA's less consciously pornographic but equally troubling "snuff" video, *Meet Your Meat*, runs aground on a Manichean split between those who work in the meat industry or eat meat, who are positioned as soulless victimizers, and the animals, who are depicted in scene after gut-wrenching scene as helpless victims.[18] Several of my students have reported becoming temporarily "converted" to veganism after watching *Meet Your Meat*, only to backslide into a guilty carnivory because they could not imagine a middle ground between the dystopia represented in *Meet Your Meat* and the vegan utopia conjured up by PETA's GoVeg.com site. I hasten to add that this critique should not be interpreted as any kind of defense of factory farming or the meat industry, or as a quarrel with PETA's philosophies. Rather, I hope that it will contribute to dialogue on the ways in which PETA and like-minded groups can more effectively *represent*, visually and thus politically, nonhuman animals and the larger environment.

One does not have to look far to find models of what I would consider a less ecopornographic politics of representation, even if these models cannot fully escape some of the same problems of commodification, complacency, violence, and so on to which mainstream ecoporn is prone. Steve Baker has studied a wide range of these models, including the revolutionary work of the duo Olly [Williams] and Suzi [Winstanley] and that of another British artist, Damien Hirst. Olly and Suzi do not simply paint animals; they expose themselves and their paintings to the interventions of nonhuman animals in the wild—anacondas, tarantulas, wild dogs, great white sharks, wolves, leopard seals—in a process that strives for "collaborative, mutual response[s]" not just on the two artists' part but on the animals'.[19] Damien Hirst creates installations using rotting cow's heads and animal parts preserved in formaldehyde; in Terry Tempest Williams's words, Hirst makes us "take note of the meat that we eat" instead of going "without thinking about the topography of the body, the cow's body, our body [. . .]" (2000, 66–67). In Williams's own collaborative work with painter Mary Frank in the books *Desert Quartet* (1995) and *The Open Space of Democracy* (2004), she resists ecopornography's efforts to conceal, consume, and objectify by joining Frank in a far-reaching dialogue, in word and image, on the nature of eros and of dialogue itself: between people of different political persuasions, between women and nature, between body and earth. Recent PBS *Nature* broadcasts such as *Shark Mountain* support the idea that popular audiences can be trusted to find educational and entertainment value in something more than the ecopornographic standard of, in the words of poet Ann Fisher-Wirth, "one mating act per meal" (2003, l. 1). *Shark Mountain* brings in a female filmmaker's voice, employs a split-screen format to show the filmmakers at work alongside the images they are obtaining, and explores the enormous costs of our hunger for seafood without casting the blame solely on nonwhite "poachers." Even MTV may hold some promise. Cynthia Chris's work on MTV's *Jackass* spin-off *Wildboyz* suggests to me that the time may be ripe for parodies of ecopornography that use subversive humor to raise ecological consciousness while still managing to skewer human greed, waste, and self-centeredness.[20] *Pace* Michael Shellenberger and Ted Nordhaus, environmentalism is not dead.[21] One crucial way in which it can achieve a larger life, however, is by reconsidering how it uses ecoporn. Environmentalism needs to keep "rethinking the human place in nature"[22] and revisualizing its visual practices accordingly. In short, it needs to join artists such as Linda

Hogan, Olly and Suzi, and Terry Tempest Williams and Mary Frank in attempting to imagine a world that truly *looks back*.

Notes

1. I would like to thank Jude Allen, Karla Armbruster, George Grattan, Richard Kerridge, Thomas Lynch, Jeri Pollock, and other members of the online community of the Association for the Study of Literature and Environment (ASLE) for their thoughtful replies to my initial query about "green porn" in October 2004. They pointed me in the direction of key sources and outlined some of the major theoretical problems explored in this chapter. One of State University of New York Press's anonymous reviewers, echoing my subconscious, pressed me to acknowledge counterarguments that needed to be addressed. Ann Fisher-Wirth shared a copy of her poem *The Kingdom of the Animals* and provided a number of useful suggestions along the way. Cynthia Chris shared draft versions of her essay "Boys Gone Wild: The Animal and the Abject" and her book *Watching Wildlife* in a particularly generous and timely fashion; she and other participants in the 2004 meeting of the Society for Literature, Science, and the Arts (SLSA) played a major role in shaping my ways of thinking about these issues. I am grateful to Alyce Miller for organizing a splendid conference, "Kindred Spirits," at Indiana University in 2006, where I was able to present a completed draft of this chapter and discuss it with Carol J. Adams, Carol Freeman, Leesa Fawcett, Jonathan Balcombe, and others. Michael Lundblad, Melissa White, and Ann Dickinson deserve special thanks for many spirited conversations, amounting to a kind of crash course in ecocriticism and animal studies, during my time with them at the University of Virginia; this chapter really began life as part of a graduate conference roundtable discussion with Mike, Melissa, and Ann in April 2002. The University of North Florida provided crucial travel and research funding for the project, and many of my students and colleagues at the University of North Florida helped road test the ideas explored here. Bill Slaughter, Alex Menocal, and Clark Lunberry are at the top of the list, but the list is long, and all have my thanks. As always, my wife Elizabeth has my eternal gratitude for her patience, honesty, and love.

2. Jessica Leigh Durfee does just this, comparing Shell's visual rhetoric with the Sierra Club's in her essay "Images of Nature in Organizational Stakeholder Publications: The Rhetorical Function of Ecopornography in Green-Washing."

3. This alternative type of "ecoporn" does exist and may be spreading; in his article "Eco-porn: Great Sex for a Good Cause," Gregory Dicum reports that one group has started using pornography (often featuring woodland settings and sexual behaviors involving plants) to raise money for rain forest protection.

4. Bill McKibben offers a disturbing overview of some of these methods in "Curbing Nature's Paparazzi."

5. I am painfully aware that, on some very literal levels, this chapter itself contributes to these problems, whether or not it is printed on recycled paper.

But I have to believe both that books such as *Ecosee* (and *Walden* and *Silent Spring* and *Refuge . . .*) will always be worth publishing, and that the paper and publishing industries will adopt more sustainable practices in response to global warming and a growing public demand for greener products. For an early preview of the future of publishing, see William McDonough and Michael Braungart's *Cradle to Cradle: Remaking the Way We Make Things*, which is printed on paper made from a fully recyclable polymer blend rather than wood pulp, and which envisions a future book that "celebrates its materials rather than apologizing for them"; this future book "tell[s] a story within the very molecules of its pages" not of "damage and despair, but . . . of abundance and renewal, human creativity and possibility" (2002, 71). To my knowledge, Andrew Ross is one of the few critics to have taken seriously the question of the larger ecological costs—as opposed to the animal welfare costs—of representing nature visually. While McKibben notes, "It is no great exaggeration to say that dolphin-safe tuna flows directly from the barrel of a Canon, that without Kodak there'd be no Endangered Species Act" (1997, 19), Ross reminds us that photography and film's contributions to environmental causes must be weighed against the fact that Eastman Kodak, for one, has traditionally been one of the worst corporate polluters in the United States. Ross describes the "chemical underpinnings of the film economy" as a whole as "sordid." It will be interesting to see how the digital revolution in photography and film will translate to a greener future for these industries too. See Ross, "The Ecology of Images" (1994, 196).

6. See Kappeler (1986, 219).

7. Although I considered removing these lines after Irwin's untimely death on the Great Barrier Reef in 2006, I left them unchanged, trusting that my critique of Irwin's show would not be interpreted as disrespect for the dead; furthermore, the manner in which Irwin died—killed by a stingray that apparently objected to being filmed at such close range—raises issues that simply must be addressed in any discussion of ecoporn. Still, it should be noted that the larger question that haunts this chapter—Does ecoporn really do more harm than good?—was posed, in so many words, by millions of people around the world in the days after Irwin's death, with many, such as Jacques Cousteau's grandson, Philippe (who was working with Irwin on the project *Ocean's Deadliest* at the time of the accident), arguing that Irwin's hands-on style awakened respect for the natural world in viewers who were normally apathetic to it (see "Cousteau: Irwin 'A Remarkable Individual' "). Philippe's uncle, Jean-Michel, gave voice to a less popular but, I believe, more ethically sound position: that it is better not to "mess with nature," as Irwin did (see "Irwin Interfered with Nature, Says Cousteau").

8. Perhaps this point will seem overly captious. After all, settlement money taken from giant tobacco companies is used to fund antismoking programs; there may be a kind of homeopathic wisdom involved in turning a tool of destruction against itself in this way. My concern, at any rate, is not so much with the license plates as with the consumerist complacency that ecoporn helps engender. For the record, the state of Florida has reported that, as of 2004, 1,385,107 of the panther license plates had been sold since the plate's introduction in 1990.

"Protect the Panther" was followed closely in the 2004 rankings by "Protect Wild Dolphins" in second place, "Save the Manatee" in fourth, and "Helping Sea Turtles Survive" in fifth. See Florida Department of Highway Safety and Motor Vehicles 2005).

9. See Haraway's (2003, 100). Haraway's book *The Companion Species,* deals with the unique relationship between humans and dogs, but it is worth asking whether a heavily "managed" animal such as the Florida panther (medicated, radio-tagged, selectively bred with other puma subspecies, and otherwise cared for much like a pet, even if it is not purposely domesticated) does not also become a kind of companion species to humans. Linda Hogan's (1998) *Power* explores another, more sustainable, type of companion-species-ship.

10. These observations on big cats can be applied, *mutatis mutandis,* to images of other endangered species. Visitors to the home page of the Bushmeat Project, for instance, encounter a photograph of a gorilla with the words "Look at My Eyes . . . Listen to Me . . . For I and my kind will be gone soon . . . Look . . . Listen . . . Help . . ." superimposed on the animal's face. Every few seconds a red target symbol flashes over the gorilla's eye. A similar image serves as the logo of the Bushmeat Project.

11. For a vastly more comprehensive and nuanced approach to televisual representations of animal sex, particularly regarding the question of how they have changed through time (due, in large part, to such scientific paradigm shifts as the rise of sociobiology, see chapter 4 of Cynthia Chris's (2006) *Watching Wildlife.*

12. For that matter, why did my students blush and turn away from copulating mandrills during a recent zoo visit, despite the concrete barriers and moats dividing, and ostensibly protecting, "us" from "them"?

13. Of course there are important technical differences between ecoporn and standard porn (shot length, distance of camera from subjects, and so on [see Chris 2006, 130–31]) as well as biological distinctions (differences in mating habits between humans and wolves and between wolves and spiders) that need to be factored in too. But I believe that these differences are ultimately less central than the reciprocities between ecoporn and standard porn, especially given wildlife filmmakers' penchant for imposing masculinist, anthropocentric narratives on their footage of nonhuman animals.

14. One major drawback of the Animal Liberation Project is that it does not address the ethical problems involved, for instance, in reproducing photographs of early-twentieth-century lynchings that were produced by white supremacists for the purpose (as various scholars have recently noted) of animalizing African Americans and schooling whites in racist ways of seeing blacks. For an extended example of the same strategy that does take note of these problems, even in its title, see Marjorie Spiegel's (1996) *The Dreaded Comparison: Human and Animal Slavery.*

15. Other noteworthy PETA projects include the work of the Lettuce Ladies, who host PETA events wearing strategically placed lettuce leaves, and a 1970's porn-style 2004 Super Bowl ad (rejected by CBS) in which two women try to seduce a pizza delivery man, only to discover that his consumption of

meat has adversely affected his "sausage." (The caption to another PETA ad reads "Eating meat got you down? Get it up. Go vegetarian!") PETA has cultivated a particularly close relationship with Pamela Anderson and other *Baywatch* alumnae (including Nolin).

16. See esp. Griffin's section "The Animal" (1981, 24–35) in *Pornography and Silence*.

17. Ingrid Newkirk, PETA's cofounder and president, has answered the charge of sexism in this way: "We don't allow sexism. We use sex. Not sexism. Nudity is not synonymous with sexism. That's an Islamic idea, a prudish idea, a sour idea" (Newkirk 2001). By invoking the idea of an abstract, gender-free "nudity," Newkirk sidesteps the thorny issue of how *female* nudity is coded in Western visual culture, and of how PETA has adapted the conventions of pornography to grab the public's attention. Her essentialist reference to Islam seems telling in this context.

18. PETA obviously has an interest in forcing the public to consider the plight of the billions of animals that *are* killed each year. Likewise, it has a clear stake in demystifying the processes by which animals are transformed imaginatively as well as physically into "meat"—from cow to "beef," from pig to "pork," and so on (see Fudge 2002, 34–46). The problem is that *Meet Your Meat*, like so many other PETA productions, does not escape the dualistic ways of thinking about the human relationship with the more-than-human world that actually has supported this massive destruction. These dualisms lead, in turn, to major flaws in PETA's handling of evidence, which does not help its cause. Not all factory farm and slaughterhouse workers take delight in *abusing* animals, for instance. In 2005, PETA actually singled out Temple Grandin, the designer of a large percentage of more humane livestock handling facilities in North America and beyond, for a "Proggy" (short for "progress") award; Ingrid Newkirk has said that "Temple Grandin has done more to reduce suffering in the world than any other person who has ever lived" (qtd. in Specter 2003, 67). More of PETA's public outreach efforts need to account for the complexities of our paradoxical relationships with other animals. For a short antifactory farming (but not antimeat) film that achieves greater rhetorical success than *Meet Your Meat* without losing sight of the vast costs to animals—and humans—of the reigning system, see the popular Internet cartoon *The Meatrix*.

19. Steve Baker observes, "Once marked by the animal, these pieces are described by [Olly and Suzi] as 'a genuine artifact of the event,' and are intended above all to bring home the truth and immediacy of these animals' precarious existence to a Western audience which has grown largely indifferent to the question of endangered species. For there to be an animal-made mark, the animal has to be present, and has to participate actively (if unwittingly). What is performed through its presence and recorded in its marks is precisely that animal's reality" (in Rothfels 2002, 88–89).

20. Chris notes that *Wildboyz* reunites several of the figures responsible for *Jackass* but "abandons the urban and suburban American settings that served as sites for *Jackass* stunts" in favor of "rainforests, outbacks, and savannahs of foreign lands where [the "Wild Boys"] undertake exercises of controlled risk

and willing abjection culled from both culture and nature" (forthcoming). In one African stunt, for instance, Wild Boys Steve-O and Chris Pontius dress up in a zebra costume and are attacked by lions. Stunts are frequently introduced by a plummy David Attenborough-like narrator's voice ("The largest African carnivore, lions thrive on the abundance of herd animals found on the Mara"), and the Boys, who may be wearing pith helmets or other Steve Irwin-style accoutrements, often announce that they are exposing themselves to danger in the name of science. While the show's interspecies ethics are obviously problematic, its success in ridiculing ecopornographic conventions deserves the attention of environmental image makers.

21. Shellenberger and Nordhaus analyze environmentalism's failure to adapt its rhetoric and politics to meet the threat of global warming, but it is worth applying their observations to environmentalism's failures in other areas, especially as these are impacted by environmentalism's *visual* rhetoric.

22. See Cronon 1996. My ways of thinking about how humans represent the natural world have been impacted powerfully by his *Uncommon Ground*.

Works Cited

Adams, Carol J. 1990. *The Sexual Politics of Meat: A Feminist-Vegetarian Critical Theory*. New York: Continuum.

Adams, Carol J., and Josephine Donovan. 1995. "Introduction." In *Animals and Women: Feminist Theoretical Explorations*, ed. Adams and Donovan, 1–8. Durham, NC: Duke University Press.

Animal Liberation Exhibit. 2005. *Animal Liberation Project*. 2005. People for the Ethical Treatment of Animals (PETA). http://www.peta.org/ animal-liberation/display.asp., accessed 18 Dec. 2005.

Baker, Steve. 2002. "What Does Becoming-Animal Look Like?" In Rothfels, ed., 67–98.

Bergman, Charles. 2003. *Wild Echoes: Encounters with the Most Endangered Animals in North America*. Urbana: University of Illinois Press.

Bushmeat Project Home Page. 2004. The Biosynergy Institute/Bushmeat Project. http://bushmeat.net/, accessed 21 Dec 2005.

Cartmill, Matt. 1993. *A View to a Death in the Morning: Hunting and Nature through History*. Cambridge, MA: Harvard University Press.

Chris, Cynthia. 2006. *Watching Wildlife*. Minneapolis: University of Minnesota Press.

———. Forthcoming. "Boys Gone Wild: The Animal and the Abject." *The Drama Review*.

"Cousteau: Irwin 'A Remarkable Individual.' " 2006. http://www.cnn. com/2006/SHOWBIZ/TV/09/06/cnna.cousteau/index.html, accessed 11 May 2007.

Cronon, William, ed. 1996. *Uncommon Ground: Rethinking the Human Place in Nature*. New York: W. W. Norton.

Crowther, Barbara. 1995. "Towards a Feminist Critique of Television Natural History Programmes." In *Feminist Subjects, Multi-Media: Cultural Methodologies*, ed. Penny Florence and Dee Reynolds, 127–46. Manchester: Manchester University Press.

Dicum, Gregory. "Eco-porn: Great Sex for a Good Cause." http://www.sfgate.com/cgibin/article.cgi?f=/g/a/2005/04/13/ gree.DTL&type=printable, accessed 21 Dec. 2005.

Durfee, Jessica Leigh. 2002. "Images of Nature in Organizational Stakeholder Publications: The Rhetorical Function of Ecopornography in Green-Washing." *Department of Communication Course Workspaces.* Dept. of Communication, University of Utah. http://www.hum.utah.edu/communication/classes/fa02/1600-1/ecoporn.pdf, accessed 21 Dec. 2005.

Durfee, Jessica Lee. 2002. "Images of Nature in Organizational Stakeholder Publications: The Rhetorical Function of Ecopornography in Green-Washing." Unpublished essay.

"Exhibit: Advertising the Environment." 1972. *Communication Arts* 14:2: 56–61.

"Famous Fur Foes." 2005. PETA (People for the Ethical Treatment of Animals). 2003. http://www.furisdead.com/FurFoes.asp, accessed 18 Dec. 2005.

Fisher-Wirth, Ann. 2003. "The Kingdom of the Animals." *Olaus Magnus,* Carta Marina. *The Malahat Review* (June):70–100.

Florida Department of Highway Safety and Motor Vehicles. 2005. "2004 Specialty Plate Sales Rankings." *Specialty License Plates.* http://www.hsmv.state.fl.us/specialtytags/SLP.html, accessed 18 Dec. 2005.

Florida Fish and Wildlife Conservation Commission. 2005. http://www.myfwc.com/panther/, accessed 21 Dec. 2005.

"Florida Panther." 2005. *Southeastern Natural Resource Center.* http://www.nwf.org/resourcelibrary/details.cfm, accessed 22 Dec. 2005.

Fudge, Erica. 2002. *Animal.* London: Reaktion Books.

Griffin, Susan. 1981. *Pornography and Silence: Culture's Revenge against Nature.* New York: Harper & Row.

Haraway, Donna. 1991. "Situated Knowledges: The Science Question in Feminism and the Privilege of Partial Perspective." In *Simians, Cyborgs, and Women: The Reinvention of Nature*, 183–201. New York: Routledge.

———. 2003. *The Companion Species Manifesto: Dogs, People, and Significant Otherness.* Paradigm 8. Chicago: Prickly Paradigm Press.

Hogan, Linda. 1998. *Power.* New York: W. W. Norton.

"Irwin Interfered with Nature, Says Cousteau." 2006. *The Sydney Morning Herald.* http://www.smh.com.au/news/world/irwin-interfered-with-nature-says-cousteau/2006/09/20/1158431752163.html, accessed September 20.

Isenberg, Andrew C. 2002. "The Moral Ecology of Wildlife." In Rothfels, ed., 48–64.

Kappeler, Susanne. 1986. *The Pornography of Representation.* Minneapolis: University of Minnesota Press.

Knighton, José. 2002. "Ecoporn and the Manipulation of Desire." In *Wild Earth: Wild Ideas for a World Out of Balance*, ed. Tom Butler, 165–71. Minneapolis: Milkweed Editions.

Kolodny, Annette. 1975. *The Lay of the Land: Metaphor as Experience and History in American Life and Letters*. Chapel Hill: University of North Carolina Press.

Mabey, Richard. 2003. "Nature's Voyeurs." *The Guardian* (March 15). http://www.books.guardian.co.uk/departments/scienceandnature/ story/0,6000,914227,00.html, accessed 11 Oct. 2005.

Mander, Jerry. 1972. "EcoPornography: One Year and Nearly a Billion Dollars Later Advertising Owns Ecology." *Communication Arts* 14:2: 45–55.

McClintock, Anne. 1995. *Imperial Leather: Race, Gender, and Sexuality in the Colonial Contest*. New York: Routledge.

McDonough, William, and Michael Braungart. 2002. *Cradle to Cradle: Remaking the Way We Make Things*. New York: North Point Press.

McKibben, Bill. 1997. "Curbing Nature's Paparazzi." *Harper's* (November): 19–24.

The Meatrix. 2003. Dir. Louis Fox. Free Range Graphics/GRACE (Global Resource Action Center for the Environment). http://www.themeatrix. com/, accessed 24 Dec. 2005.

Meet Your Meat. 2003. PETA (People for the Ethical Treatment of Animals). http://www.goveg.com/factoryFarming.asp, accessed 24 Dec. 2005.

Millet, Lydia. 2004. "Die, Baby Harp Seal!" In *Naked: Writers Uncover the Way We Live on Earth*, ed. Susan Zakin, 146–50. New York: Thunder's Mouth Press.

Mitman, Gregg. 1999. *Reel Nature: America's Romance with Wildlife on Film*. Cambridge, MA: Harvard University Press.

Newkirk, Ingrid. 2001. "Ingrid Newkirk: Taking On the Critics." Interview. *Animal Liberation*. Animal Liberation New South Wales. http://www. animal-lib.org.au/activists/activists/newkirk-ingrid.htm, accessed 21 Dec. 2005.

Ross, Andrew. 1994. "The Ecology of Images." In *Eloquent Obsessions: Writing Cultural Criticism*, ed. Marianna Torgovnick, 185–207. Durham, NC: Duke University Press.

Rothfels, Nigel, ed. 2002. *Representing Animals*. Bloomington: Indiana University Press.

Shark Mountain. 2005. Dir. Howard Hall. *Nature*. (October 16) PBS. WNET, New York.

Shellenberger, Michael, and Ted Nordhaus. 2004. "The Death of Environmentalism: Global Warming Politics in a Post-Environmental World." In *Welcome to the Breakthrough Institute*. http://www.thebreakthrough. org/index.php, accessed 30 Dec. 2005.

Specter, Michael. 2003. "The Extremist." *The New Yorker* (April 14): 52–67.

Spiegel, Marjorie. 1996. *The Dreaded Comparison: Human and Animal Slavery*. Rev. ed. New York: Mirror Books/I.D.E.A.

Williams, Joy. 2005. "Save the Whales, Screw the Shrimp." In *Saving Place: An Ecocomposition Reader*, ed. Sidney I. Dobrin, 203–13. Boston, MA: McGraw-Hill.

Williams, Terry Tempest. 2000. *Leap*. New York: Pantheon Books.

Williams, Terry Tempest, and Mary Frank. 1995. *Desert Quartet*. New York: Pantheon Books.

———. 2004. *The Open Space of Democracy*. New Patriotism 4. Great Barrington, MA: The Orion Society.

3

Ecology, Images, and Scripto-Visual Rhetoric

Heather Dawkins

In the social history of art that has flourished over the last three decades, the word "rhetoric" is not often used. Indeed, contemporary use of the word in the visual arts is largely in reference to its debased and derisive meaning. The art world is, to generalize, suspicious of rhetoric, whether written or visual. Such images and texts are found to be conventional, manipulative, and saturated with power. As a result, the kinds of images art historians tend to work on, and the kinds of images artists are encouraged to create, are deliberately antirhetorical, open in structure, meaning, and resonance. Nevertheless, images continue to be used in many forms of persuasive communication outside of the fine and experimental arts, and art history could have much to contribute to the study of such imagery, should art historians choose to analyze it.

In the social history of art that emerged after 1970, images were examined for their ability to represent meanings that had been socially and historically suppressed, to disrupt ideologies, to provoke revolution, or alternatively, to buttress existing relations of power.[1] From the outset, the social history of art was deeply informed by conflicts of class, gender, or race, and by the possibility that suppressed or repressed meanings could be embodied or represented by images. The task of the social historian of art was to research and tease out the multiple and complex meanings that often were obscured by the cultural reverence for art, artists, and creativity, and to analyze the relationship between images and the dynamic process of hegemony.

In the 1990s, the preoccupations of social historians of art shifted somewhat, with less emphasis on the role of images in relation to structures of power and more emphasis on the instability and uncertainty of images, perception, and the production of meaning. With the insights of poststructuralism, deconstruction, and Peircean semiotics, the meaning of images began to be seen as a complex, multilayered, fluctuating negotiation between viewers and situated knowledges, between perceptual intuitions and shared aesthetic frameworks, and between signifiers and semiotic processes. Studying the multilayered complexity of the image sometimes led to a consensus about the meaning and significance of the image to a certain group of people in a specific time and place. But more often, it led to an awareness of the fragility of meaning in relation to images and to an emphasis on the instability of the perceptual processes through which images are grasped.[2]

In the social history of art of the last three decades, the word rhetoric is little used, yet much of the analysis that art historians undertake overlaps with the analysis of rhetoric. Art historians analyze how images produce perceptions, meaning, and affect for a viewer or an audience. They study how images are composed from multiple, intertextual elements of public and private symbols, from visual forms, and through creative practices. They examine how images position viewers perceptually and spatially—as well as within a matrix of social meaning and aesthetic resonance. They analyze how images work, or not, and for whom. These are all elements in the analysis of rhetoric. There are strong parallels between the analytical aims of the social history of art and rhetorical studies.

Although there are common aims in the social history of art and rhetorical studies, this does not hold for art images and rhetorical texts as such—here there are more disjunctions than parallels. Rhetorical texts are a long-standing tradition in, and dynamic contemporary form of, literary and oral cultures; their contemporary characteristics are to define the good, to seek through a variety of means effective persuasion, to create a community of listeners, and to propose a course of action to them. Many images also are rhetorical, but they feature most prominently in advertising and mass media. This is the source of a chasm between art history and rhetorical studies.

Art history has been shaped in part by the history of modern art, which is replete with forms of art making that do not intend to reason, persuade or convince, propose the good, or encourage a course of action—although these modern art forms have indeed shaped a community of viewers. Many modern artworks illustrate nothing and refuse to communicate. They take their "thingness" seriously and refuse to extend themselves to the viewer. If the study of rhetoric is a study of meanings crafted from language held in common, of persuasive communication used to create cooperation and community, then much of the art of modernity is a conscientious objector. Dismissive of, or opposed to, visual

language held in common—the instrument of mass media and aesthetic
banality—experimental artists, critics, and curators sought (and continue
to seek) images that are innovative, unfamiliar, individualistic, difficult,
or opaque. Some of these images may have much to say about nature
or ecology, but the mode in which they address viewers is deliberately
antirhetorical.[3] Art history has been profoundly shaped by this trajectory
in modern art. It developed as a discipline during the nineteenth and
twentieth centuries, a time when the most visually sophisticated art of
Europe was freed from its role as rhetoric and propaganda for ruling
elites and was simultaneously defined in opposition to emerging forms
of mass communication.

   All art may somehow affect the world, play a role in change, or
lend itself to unforeseen ends, but if modern art has played a role in
change, then it has done so while eschewing the structures of rhetoric.
Yet the images and traditions of art are now incorporated into many
forms of rhetorical address—rhetoric crafted by corporations, social
movements, faith communities, advertisers, leaders, citizens, and activists.
How do the images, traditions, or contemporary forms of art function
within the inherently linguistic structures of rhetoric? How do images
function within the specific tradition of environmental rhetoric, or within
the specific problematic of the environment?

   I approach these questions by analyzing two projects from the
period 2004–2005. Both are imbricated in mass media but are a step
removed from advertising for profit. One is the Greenpeace calendar for
2005; the other is an international exhibition and pictorial book called
Massive Change.[4] Their rhetorical structures and use of imagery are distinct and differ in historical, aesthetic, semiotic, and affective complexity.
In both cases, the meanings, significance, and effects of images emerge
from the interaction between pictorial structures and language, that is,
from a scripto-visual matrix.

                              I.

Scripto-visual ecological rhetoric is well established in print culture, where
leaflets, calendars, brochures, and illustrated books have finely crafted
conventions for combining word and image in purposeful rhetorical
address. A quintessential recent example is the 2005 calendar "Greenpeace: Stepping Lightly on the Earth."[5] The calendar features 12 x 15"
color photographs, in saturated hues, of landscapes, flora, and fauna.
These genres have a pedigree that goes back centuries, but they were
most intensively developed in the late eighteenth century as Europeans

sought to explore and represent scientifically the regions of the world. The Greenpeace calendar relies on the familiarity and stability of these modes—landscape, flora and fauna—to create an emotional resonance in viewers and to arouse feelings of awe and wonder in relation to nature. Significantly, the calendar starts and ends in the mode of the sublime, an aesthetic that was formatively defined in the eighteenth century as a secular experience of transcendence, a feeling of being overawed while losing a sense of oneself.[6] Indeed, the January image is a photograph of dawn breaking on the peaks of the Karakoram Mountains in Pakistan, with the spectator positioned at the center of a dark, narrow ledge. It is a classic invocation of nature as awe inspiring, transcendent, and luminous and of "man" as a puny and precarious onlooker.

The calendar ends in a similar mode of spectatorship, with December illustrated by a picture of ice flows in the Weddell Sea, Antarctica. The ice flows are monumental, an intense, translucent cerulean to cobalt blue, and strangely biomorphic. They dwarf the viewer, who is implicitly positioned in suspension above an icy sea. A flock of penguins on the nearest ice flow indicates size, but the blue ice flows beyond rise in incomprehensible scale and magnitude. The December illustration constructs an experience of the sublime; unlike the January image, which could have been made by other means in the eighteenth or nineteenth centuries, the December image could only have been made at a time when technology succeeded in withstanding the Antarctic's extreme conditions.

The Greenpeace calendar thus opens and closes by establishing an affective, somatic resonance for viewers, using pictorial conventions of the sublime established during the eighteenth century. Those conventions destabilize the spectator as a bounded subjectivity, diminish the spectator in proportion to the perceptual field, and compel a secular-religious, affective intensity—a sensation of being overawed by the grandeur and wonder of the natural world.

The remainder of the Greenpeace calendar also uses historical genres, although more intimate ones. Three of the months feature landscapes, but they are closer to the pastoral than sublime, with each one foregrounding flora as a predominant focus.[7] Faunas dominate the remaining six months. Each is illustrated with a different species: polar bears, a black-browed albatross, and sea lions; an angelfish, jaguar, and sea otter. The images of animals work to establish an emotional resonance with viewers, using interspecies affinity, zoological curiosity, or anthropomorphic parallelism.

In a border of short texts and small photographs below each calendar grid, there is another strategy at work. On the left is a text that is naturalistic, ecological, and pedagogical. It outlines the distinctive features of the landscape, flora or fauna for the month, describing, for example, the migration of polar bears, coral ecosystems, sea otter populations, or jaguar behavior. These texts describe distinctive features of the species and their habitat. The brief text (about 100 words) is mirrored on the lower right with slightly longer descriptions of Greenpeace activities related to the place, flora, or fauna for the month. Greenpeace campaigns against genetically engineered crops, overfishing, nuclear energy, habitat loss, or persistent organic pollutants are described. Framing both texts—the pedagogical on the left and the activist on the right—are small documentary photographs in color, of protests, green technologies, Greenpeace activism, and environmental degradation. Thus the viewer/reader is led, step-by-step, from a vivid, pristine, emotionally resonant image (the large feature photograph for the month), to learning about species and habitat, to anthropological threats to nature, and finally to Greenpeace activism, which the very purchase of the calendar has supported.

The calendar layout for April demonstrates the typical stages of progression in this rhetorical structure, illustrated with a photograph of three black-browed albatross. In front is a downy grey chick staring out at the viewer; behind it, its parents, with their vibrant cadmium orange to lemon yellow beaks, black swooping eyebrows, and black wings on pure white bodies. The parent on the right has its head bowed to look at its young. Below this detailed, strikingly anthropomorphic image is the calendar grid and below it the pedagogical text. It is titled "Nuclear Family," and it explains:

> The Malvinas, or Falkland Islands, in the South Atlantic are home to 85 percent of the world's population of black-browed albatross. The bird, with a wingspan of over seven feet, is one of nature's true long-haul fliers. During its lifetime, the albatross travels hundreds of miles all over the Indian Ocean, South Atlantic, and Southern Ocean. During a foraging expedition, albatross skim the waves taking food from just below the surface. Black-browed albatross generally have the same mate for life and produce only one chick each year. The egg hatches in October and the young bird is fed by both parents until it flies the nest in March.[8]

This account of species characteristics and familial fidelity is complemented on the right with a short description of nuclear danger to the oceans. Greenpeace's oceangoing antinuclear protests are represented in this text and documented in three small (under 1.5 x 2") photographs that frame the pedagogical and the environmental text. The environmental complement for the pedagogical text, quoted earlier, reads:

> While albatross ride the air currents above the oceans, nuclear transport ships navigate below with their deadly cargo. Spent weapons-grade plutonium is carried by ship between France, Britain, and Japan across some of the most dangerous seas in the world. These high-level waste shipments contain more radioactivity than was released at Chernobyl and enough material to make 50 nuclear bombs. Countries such as Argentina have demanded a halt because of the dangers posed by accident or terrorist attack. Nuclear waste is produced at every stage of the fuel cycle, from uranium mining to reactors to reprocessing. Greenpeace wants governments to shut down all nuclear plants, stop building new ones, and invest in clean, renewable energy from wind, wave, and sun.[9]

The calendar layout takes the viewer through the stages of increasing ecological awareness, culminating in Greenpeace activism. Nevertheless, through sheer size and color saturation, it is the family of albatross that dominates the perceptions of the viewer/reader. The pleasures of that image are conventional and reassuring, and its familiar but resonant harmony is a counterweight for the extreme dangers posed by nuclear energy and nuclear waste.

The pictorial genres used in the Greenpeace calendar—sublime and pastoral landscapes, flora and fauna, documentary—are typical of contemporary environmental rhetoric. Greenpeace's structure of image and text is particularly appealing to urban spectators, who can be drawn to the tranquility and beauty of nonurban spaces and to the unexpected otherness of nature while also being reminded of the countless threats to nature by industry, development, tourism, or anthropological climate change. In this delicate balance, the deep structures and familiarity of pictorial modes developed in the eighteenth century are crucial. They contribute stability and continuity, they compensate for the descriptions of modernity's unremitting threat to the ecosystems of the planet, and they prevent the calendar from becoming apocalyptic.

II.

The significance of stability in the scripto-visual rhetoric as crafted by Greenpeace can perhaps best be understood by comparison with a project that also is scripto-visual, rhetorical, ecological, and international yet lacks the calendar's deep structures of historical continuity and its graduated rhetorical trajectories. The project is *Massive Change*, which began circulating internationally in 2004 as an exhibition and a book.[10] The exhibition was first conceived as being about contemporary design, but it developed into a fearlessly rhetorical project—I say "fearlessly" because most art galleries avoid overtly rhetorical exhibitions. The premise of the exhibition is that design is not about style or visual phenomena but is rather about the multitude of decisions shaping the world. As the exhibition and book proclaim, *Massive Change* "is not about the world of design; it's about the design of the world."[11] In "the design of the world" *Massive Change* identifies sustainability as the central concern, essential to problem solving for water, housing, poverty reduction, transportation, energy, genetic engineering, recycling, and so on. The issues of ecology and environment are implicit throughout the project as well as being explicitly stated at regular intervals.

Massive Change is, above all, constituted through images. The exhibition and the book are composed of a multitude of images and objects documenting projects, materials, and inventions that have current and future potential for creating a sustainable world. The book teems with image intensity—there are some 1,044 within its 245 pages. Even pages that are primarily textual contribute to the visual intensity of the book. Within these pages, selections of the text are magnified (printed in dramatically oversized fonts), creating both nonlinear eye movements and dramatic shifts in perceptual experience within one page.[12] Primarily textual, double-spread pages of interviews with social entrepreneurs, activists, and researchers are radiant yellow, suddenly accentuating the book and its pages as hand-held objects rather than transparent texts. And then there are the images—water purification systems, green roofs, solar energy, public transit, fuel-efficient stoves, biological organisms, green vehicles, wind power, micro and nano structures, engineered tissues and organs, materials that are renewable, super insulating, self-repairing, or biodegradable, particle accelerators, graphs, maps, super computers, power accelerators, and solar cells. There are images made by cameras, computers, electron microscopes, satellites, ultrasound, radiation, field-ion microscopes, telescopes, and radar. These diverse images are organized by topic—transportation, housing, energy, or materials engineering, for example. The images also are drawn into a simple, repetitive, and percussive textual structure that forms the call to action in rhetoric. Each set of images and objects is introduced by bold declarations in oversize lettering that are positioned in the exhibition at the beginning of each display and in the book on the first page of each chapter. They proclaim, "We will enable sustainable mobility," "We will bring energy to the entire world," "We will eliminate the need for raw material and banish all waste," and "We will eradicate poverty."[13]

The "We" in these statements is an abstraction, an ideal community of "change makers" of the early twenty-first century, consciously pursuing technological, social, and ecological change through multifaceted research and design (in the expanded sense of designing the world).[14] The profusion of examples of research and design creates a factual texture that is overwhelming and potentially bewildering, but the cacophonous fragments of information are punctuated and contained throughout the monolithic, inclusive pronouncements. The viewer is propelled through the exhibition and the reader through the chapters of the book, with optimism and purposeful determination.

The images have a different role, slowing down the viewer and positioning him or her to consider the material world and human ingenuity in shaping and transforming it. The images invite viewers to consider the creativity behind the diversity of research and invention, and when the knowledge of the viewer limits comprehension, the images sustain perceptual and aesthetic contemplation. The images are an intricately detailed montage, background to a definition of "the good" as human initiative and ingenuity used to create sustainable systems and improve the welfare of the whole human race. This goal may sound accommodating of modernist rationalization, centralization, or bureaucracy, but *Massive Change* is resolutely postmodernist. Embracing postmodern plurality and globalism, the projects included in *Massive Change* are decentralized geographically, diverse technologically, and on different orders of creativity and method. The project features high and low technologies, scientific and mechanical experiments, and biological and genetic processes; it integrates urban planning, social theory, development issues, and economics. The change makers included in the project are equally diverse—individuals, community groups, NGOs, university and scientific researchers, corporations, social entrepreneurs, and volunteers.

In *Massive Change*, the modernist belief in innovation and progress is conjoined with postmodern plurality, fragmentation, diversity, interdisciplinarity, and globalization. The essential experiential mode of postmodernism, visuality, provides a solution to the problem of integrating highly specialized research in the larger context of a common language conducive to cooperation and community. Images are used to represent scientific processes that are unfathomable to the layperson; the images readily lend themselves to untutored perception. Indeed, I was surprised to see how many images on a microscopic, genetic, or nanoscopic scale echo the history of abstraction or familiar astronomical imagery of the cosmos.

Massive Change is founded in modernist axioms of progress and innovation as well as postmodern plurality; it is historically nonsynchronous. This nonsynchronicity is the basis for the most important cultural feature of the exhibition and book—their iteration of eco-optimism. The book and exhibition amass scores of examples of human ingenuity in research, design, architecture, science, activism, business, government, urban planning, corporate capitalism, and technology, orchestrating them as the groundswell of change that will result in both sustainability and global equality. Images play a central role in the support for eco-optimism, as they make up the pluralized and decentralized foundation for human ingenuity. The shape that ingenuity takes is often highly specialized, often technological, and, occasionally, environmentally surprising.

Anecdotally, it was apparent that some of the 60,000 visitors to the exhibition at the Vancouver Art Gallery were uneasy in seeing genetically modified plants and animals alongside technologies designed to reduce the impact of humans on the environment.[15] Others found themselves uneasy being propelled through the exhibition by rhetorical affirmations, and all the more so when the affirmation read, "We will design evolution," followed by examples of genetic engineering, agricultural biotechnology, tissue engineering, and transgenic technologies.[16] These research practices are more often shunned in environmental and ecological rhetoric, but they could be included in *Massive Change* because the project's visionary eco-optimism is based in a precise configuration of nature. Nature is here conceived as a complex topography of cosmic, micro, nano, and biochemical intricacy. Scientific images are the primary vehicle for the portrayal of nature as a universe within the laboratory, and in that universe, nature unfolds as infinite complexity. *Massive Change* also defines nature on a planetary and organic level, but tangentially, as a destructive force, as a productive support for urban conglomerations, and as a model for chaordic structures.[17] The absence of a conception of nature as other, or as a limited totality of interdependent parts, narrows the ecological ethic in *Massive Change* and constructs nature as an inexhaustible resource for biomimetics and materials engineering. In *Massive Change*, nature is represented through the contemporary quest for knowledge and mastery of the micro-natural world. Images and image technologies are crucial in constructing an interface between the science of the micro-natural world and the public.

Take, for example, the section of the project on imaging technologies, introduced by the statement, "We will make visible the as yet invisible."[18] This section of the book and exhibition features images from beyond the narrow range of the electromagnetic spectrum to which human eyes are sensitive. There are images made by electron microscopes, x-rays, electronic diffraction equipment, infrared cameras, MRIs, ultrasound, sonar, gamma rays, and atomic detectors. In the exhibition, these highly specialized, extraordinary images surround and overwhelm a narrow strip of color photographs collected from everyday life. The color photographs are visible in the black-and-white installation photograph (Figure 3.1) as a band of compressed rectangles in the middle ground. In the book, the amateur color photographs comprising two pages are sandwiched between an exceptional range of black-and-white images made by infrared, sonar, ultrasound, magnetic resonance, radar, radio, x-rays, gamma rays, and multispectral telescopes.[19] Whether they are cosmic, subatomic, or cellular, these images abstract and expand time and space. Matter unfolds beyond its surface with a complexity, intricacy, and mystery that is unfamiliar and destabilizing. Far from the techniques of the Greenpeace calendar, which called on interspecies affinity, anthropological parallelism, or ecological curiosity, as well as sublime and pastoral landscapes, *Massive Change* positions the spectator in wonder at scientific complexity and at the mystery of matter. Compare, for example, the familiar photographs of animals, as found in the Greenpeace calendar, to a photograph in *Massive Change* (Figure 3.2).[20] It is not recognizable as animal at all, yet it is an image

Figure 3.1. *Massive Change* is a project by Bruce Mau Design and the Institute without Boundaries, commissioned and organized by the Vancouver Art Gallery.

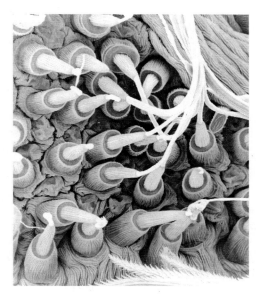

Figure 3.2. This image of spider spinnerets was generated by a scanning electron microscope. At 2.5 inches wide, the image reveals a detailed topography 973.62 times the original size.

of spider spinnerets, produced by an electron microscope. It is this kind of image making that renders nature an unlimited resource for the engineering of materials, structures, and life-forms and as a universe within the laboratory. The emphasis throughout the book and exhibition is on the new, ever-expanding possibilities of invention that can arise from understanding micro-complexity on a planet constituted through, and saturated with, largely unknown natural processes.

The optimism of *Massive Change* is undeterred throughout the exhibition. Whether the topic is housing, technology, medicine, materials, market economies, transportation, or recycling, the world is portrayed as if radical humanitarian and ecological change is within its grasp. In the book, however, a sequence of photographs evokes the scale and difficulty of the ecological challenge. Three double-page spreads reproduce color photographs of refuse by Edward Burtynsky; in each photograph, the discarded objects fill the frame and suggest their profusion to infinity. Countless tires form a bleak landscape in the first photograph, densified oil drums form a massive wall in the second photograph, and innumerable phones are heaped up in the third.[21] The image of the tire-filled landscape is superimposed with the rhetorical call to action, "We will eliminate the need for raw material and banish all waste," but that future-oriented ideal is overwhelmed by the magnitude of waste depicted in each photograph. Whether or not they are headed for recycling, the discarded objects stand as a mute and depressing testimony to the global mass production and discard of toxic materials.

It is this depressing legacy of modernity that *Massive Change* must both assume and distance itself from in order to persuade viewers/readers of its eco-optimism. *Massive Change* sees sustainability as a question of boundless creativity, curiosity, and invention, dependent only upon the resourcefulness and innovation of individuals and groups in a diverse and pluralized world, and on the secrets of matter being made legible for manipulation or mimetics. It rules out no technology, science, or research practice in its pursuit of the good of sustainability, and it is this postmodern boundlessness that makes it unfamiliar as ecological rhetoric. Its scripto-visual rhetorical constructions may succeed with mainstream audiences and institutions, because that realm finds optimism reassuring, warranted or not. Within environmental movements, however, the eclecticism and postmodern incoherence of *Massive Change* will render it ineffective as scripto-visual ecological rhetoric.

Both the Greenpeace calendar and *Massive Change* are intended for urban audiences, although their appeal to them differs greatly. The Greenpeace calendar offers the urban viewer an image of the tranquility and otherness of nature, yet links it to environmental crisis and activism. In contrast, *Massive Change* offers an urban audience an experience that is analogous to the intense, disjointed, and disjunctive montage of urban conglomerations: an image of sustainability as the result of a postmodern, chaordic matrix of innovation. In both projects, images are essential; they are key elements in a scripto-visual rhetorical structure. The images work to define the good, to create a community of viewers and readers, to persuade, and to propose action.

III.

The distinctive role of the imagery used in the Greenpeace calendar and in *Massive Change* can perhaps best be evaluated by comparing them to an earlier, instrumental, and enduring form of scripto-visual environmental rhetoric. I am thinking of Rachel Carson's book *Silent Spring*, published in 1962.[22] The illustrations, commissioned from Lois and Louis Darling, are line drawings featured on the opening page or two of each chapter. They picture rural landscapes, rivers, and suburban neighborhoods—aerial, agricultural, and household sprays—deer, cougars, raccoons, squirrels, birds, and fish—and microorganisms, cells, and molecular structures. Unlike the intrinsically rhetorical scripto-visual strategies of the Greenpeace calendar and *Massive Change*, the images in *Silent Spring* are primarily an affective supplement to the text's rhetorical structures. The line drawings set a tone of quiet familiarity at the beginning of each chapter, using well-established pictorial conventions for landscapes, flora, fauna, or modern science. Through line and composition, the images convey a harmonious balance, and this is the case even when the image includes an otherwise discordant element, such as a pesticide sprayer.[23] The simplicity and the stylistic harmony of the images are unsettled over the course of the book, however, as each chapter reveals another aspect of the chemical interconnectedness of the biological world, and the consequences of introducing pesticides into the web of interdependence upon which life depends.

The first third of *Silent Spring* reveals the intricate complexity and balance of nature as well as the dangers of pesticide use. Those chapters are illustrated with landscapes, plants, or icons of modern scientific knowledge (for example, a molecular structure or a microscopic view of soil).[24] As the reader progresses through the book, the harmony and simplicity of the images are increasingly untrustworthy, as the text establishes the invisible presence of toxic chemicals disseminated through air, water, or soil and concentrated through the food chain. The visual harmony of the images is gradually undermined, displaced by disquiet and doubt. As the images acquire meanings through their scripto-visual context, their emotional resonance shifts from harmony to apprehension, from familiarity to anxiety. The images come to be informed by an invisible—pervasive and toxic—threat.

When considered scripto-visually, the images in *Silent Spring* are more complicated than they might first appear. Even so, they are far from the complex rhetorical structures that can be found thirty years later in the Greenpeace calendar or *Massive Change*. Removing the images from the Greenpeace calendar or from *Massive Change* would decimate their rhetorical structure and effects; removing the images from *Silent Spring* would only alter its affective resonance. Comparing the scripto-visual strategies of the Greenpeace calendar and *Massive Change* to those of *Silent Spring* reveals that ecological rhetoric now uses visual culture in complex ways. Images are still used to construct affective resonance, but they also are used to convey essential meanings, to demonstrate ideas, to define the good, to persuade, and to propose action; they have taken their place as intrinsic rhetorical structures in environmental rhetoric. One rhetorical function stands out from the rest, however. Images are now used to mediate the public's relationship to highly specialized scientific or ecological research, creating a community of viewers and readers across disparities of knowledge.

Notes

1. I use the "social history of art" to refer to scholarship that encompasses a range of methodologies—Marxist, feminist, postcolonial, psychoanalytic, poststructuralist, and so on. Despite the range of approaches used, this research is centered on an inquiry into art's social context—its production, reception, aesthetic structures, meaning, and significance. Some art historians use "the social history of art" to refer to Marxist approaches and would describe the broader field as the critical, radical, or new art history.

2. For example, a recent art historical publication examines the perceptions, meanings, and historical significance of paintings of trees by Théodore Rousseau. With theoretical and historical rigor, Greg Thomas argues that Rousseau developed a new way of representing trees, and that his creative practices and paintings are shaped by a nascent ecological sensibility. See Thomas 2000.

3. For experimental art related to nature and ecology, see: Boettger 2002; Kastner and Wallis 1998; Spaid 2002.

4. Bruce Mau Design and the Institute without Boundaries, *Massive Change: The Future of Global Design*, commissioned and organized by the Vancouver Art Gallery (October 2, 2004, to January 3, 2005). See Mau, with Leonard and the Institute without Boundaries 2004.

5. See Greenpeace 2005.

6. The theory of the sublime, elaborated by Edmund Burke in 1757, was conducive to artistic uses because he described it using primarily visual phenomena. Waterfalls, mountains, and fuscous colors were examples of the sublime. As these examples indicate, Burke's delineation of the sublime had to do with the characteristics of objects and was instrumental in the development of a pictorial genre that has persisted to this day. See Burke 1987. Kant's late-eighteenth-century reworking of the sublime shifted the emphasis to its origin in, and implications for, subjectivity. The illustrations for January and December in *Greenpeace: Stepping Lightly on the Earth 2005* are more closely related to the Burkean definition of the sublime than the Kantian. See Kant 1961.

7. Greenpeace, March, May, and August n.pag.

8. Ibid., April n.pag.

9. Ibid.

10. The exhibition opened at the Vancouver Art Gallery (October 2, 2004, to January 3, 2005). It was shown at the Art Gallery of Ontario in Toronto (March 11 to May 29, 2005), and at the Museum of Contemporary Art in Chicago (September 16 to December 31, 2006). The *Massive Change* Web site originally indicated that the exhibition would travel to additional venues in the United States, Europe, and Asia, but that information has been withdrawn.

11. Mau et al. 2004, 11.

12. Only three dozen pages have no illustrations, and all of these use typeface as a compositional element.

13. Mau et al. 2004, 46, 68, 180, 216.

14. The *Massive Change* Website defines a change maker as "a person or group who works to solve the world's greatest challenges. Change makers are not afraid of complex problems. They use their education, talents and creative will to solve complicated problems, and work on both large and small-scale projects." See http://www.massivechangeinaction.virtualmuseum.ca/tookit/glossary/htm, accessed 1 Nov. 2005.

15. The unease about biotechnology was apparent in comments, written by visitors, that were posted on a wall in the gallery and in conversations about the exhibition. Within the biotechnology display in the exhibition, viewers were invited to indicate their approval or disapproval of agricultural, genetic, or medical biotechnology by inserting a yellow slip of paper into a Plexiglas box to indicate whether they were for or against a particular type of research. The quantities of paper in each box could not be taken as an indication of collective opinion, however, as nothing prevented someone from putting a handful of votes in one box.

16. This section of the exhibition is represented by a chapter in the book. See Mau et al. 2004, 196–215.

17. According to *Massive Change,* "chaordic" is a combination of the words "chaos" and "order" and is used to describe "decentralized collaboration within a single organization." Chaordic systems are said to have a resiliency that mimics nature. *Massive Change* provides the example of Visa as a chaordic system, quoting its founder Dee Hock: "Visa has elements of Jeffersonian democracy, it has elements of the free market, of government franchising—almost every kind of organization you can think about. But it's none of them. Like the body, the brain, and the biosphere, it's largely self organizing." See Mau et al. 2004, 135.

18. Mau et al., 2004, 106–25.

19. The amateur photographs are featured on pages 116–18. They are preceded and followed by black-and-white, high-tech images that begin on page 106 and end on page 125.

20. Mau et al. 2004, 120.

21. Ibid., 180–95.

22. Carson 1962.

23. Ibid., 5, 154–55, 173.

24. Ibid., 15, 52.

Works Cited

Boettger, Suzaan. 2002. *Earthworks: Art and the Landscape of the Sixties.* Berkeley: University of California Press.

Burke, Edmund. 1987. *A Philosophical Enquiry into the Origin of Our Ideas of the Sublime and Beautiful.* Oxford: Basil Blackwell.

Carson, Rachel. 1962. *Silent Spring.* 1962. Greenwich, CT: Fawcett.

Greenpeace. 2005. *Greenpeace: Stepping Lightly on the Earth 2005.* New York: Workman Publishing.

Kant, Immanuel. 1961. *The Critique of Judgement.* Oxford: Clarendon Press.

Kastner, Jeffrey, and Brian Wallis. 1998. *Land and Environmental Art.* London: Phaidon.

Mau, Bruce, with Jennifer Leonard and the Institute without Boundaries. 2004. *Massive Change.* New York: Phaidon.

Spaid, Sue. 2002. *Ecovention: Current Art to Transform Ecologies.* Cincinnati, OH: Contemporary Arts Center, ecoartspace, and greenmuseum.org.

Thomas, Greg M. 2000. *Art and Ecology in Nineteenth-Century France: The Landscapes of Théodore Rousseau.* Princeton, NJ: Princeton University Press.

4

Field Guides to Birds

Images and Image/Text Positioned as Reference

Spencer Schaffner

Page ahead, now back. Too far—scan the index. Binoculars up, focus—look, look—return to the guide. Contemporary field guides to birds are visual compendiums of images and text designed to facilitate the speedy and accurate identification of birds by sight. Few organized outdoor pastimes can compare when it comes to the visuality of bird-watching, with binocular-enhanced bird-watchers abstracting the class (*aves*) from other living creatures and complex habitats, training magnified gazes on individual birds, and then working to classify and name birds with few exceptions. Visual bird classification as a hobby is dependent upon field guides, but the visual content of such quick-reference texts only establishes ways of seeing, encountering, and thinking about avifauna through the reference position that field guides maintain in bird-watching as a social practice. Field guides are critical to bird-watching as a socio-rhetorical activity because of this reference position.

Close attention to visual representations of birds and image/text relationships in field guides is rewarded when one understands bird-watching as a textually mediated environmental encounter producing specific ways of considering and visualizing the environment. This is because unlike some genres that attempt to control environmental encounters but are often overlooked (maps, for instance, or signage at designated "scenic areas"), field guides to birds are obsessively, and repeatedly, referred to, locating the guides in a central position amid what Carolyn Miller refers to as the "rhetorical action" of bird-watching. Contemporary field guides are visual texts for a visual pastime, and taxonomic discourse predominates in both image and text. While this is a pastime in which novice and highly experienced bird-watchers evidence a variable system of dispositions (Bourdieu 1984), including idiosyncratic tastes, preferences, and practices, it is the taxonomic discourse of contemporary field guides that is foundational to most bird-watching encounters. Prominent in bird-watching field guides, taxonomic discourse exerts itself upon bird-watching ways of seeing and interacting with avifauna because of the reference position that field guides hold. To the extent that field guides function as regulatory texts, then, it is their taxonomic authority that field guides most circulate, through image/text positioned as reference, in bird-watching as a social practice.

This dynamic genre positions and encourages reader-users to think primarily and obsessively about one rhetorical accomplishment—successful identification and naming—and while this may seem relatively apolitical and inert from an environmentalist perspective, bird-watching sustains a belief in itself as an environmentalist pastime. This is the case even though, as I will explain, textually mediated forms of bird-watching relinquish themselves to regulated forms of looking that reproduce inherently moderate environmental sensibilities. As Jeffrey Karnicky writes, "Bird-watching accommodates birds to a human visual apparatus, to allow bird-watchers to differentiate birds by species (and subspecies and population), age, and sex" (2004, 254). This accommodation, I am arguing, involves several steps, with the field guide functioning as a crucial tool in the translation.

Taxonomic Field Guide Discourse

Check Sibley, now National Geographic—still not seeing it. Google image search for "white wing bars." Scan images, that might be it. Back to Sibley. As quick-reference texts, highly designed contemporary field guides to birds enable fast, easy, and repeatable identification involving habituated and preestablished sequences of steps. Within the quick-reference for-mat—which involves images and species descriptions aligned in tidy page sets; predictably placed content on every page to enable flipping and scanning; searchable indexes; and thumb-reference features such as icons placed in page corners and color-coded page edges—taxonomic discourse predominates, defining the nature and configuration of both image and text. Contemporary field guide images can be defined as taxonomic, as opposed to more naturalistic and anthropomorphic images found in some nineteenth-century guides (Wright 1895; Blanchan 1887/1897/1932), because contemporary images aim to represent defining field marks that can be used by bird-watchers in the field for purposes of differentiat-ing similar-looking species; descriptions can be defined as taxonomic because they list these features, adding characteristics that are not easily represented visually (such as behavior, habitat, call, and song). It is the pervasiveness of taxonomic discourse, informing nearly all aspects of image and text in contemporary field guides, that puts pressure on the specific formation of bird-watching dispositions. In this formation, or sociality, identifying and naming birds is a *first act* in any bird-watching encounter. Discourses of romanticism, transcendentalism, aestheticism, and anthropomorphism may inform bird-watching encounters, but only after the first act of identification, based on taxonomies of distinguishing traits, has been accomplished. And when this first act cannot be fulfilled, then quick-reference work takes place, either on the spot or after the fact. It is the pervasiveness of taxonomic discourse in contemporary field guides that promotes bird-watching as a distinguishing sensibility.

Michael Lynch and John Law, in their article "Pictures, Texts, and Objects: The Literary Language Game of Bird-Watching" (1999), describe differences in field guide images (in terms of modality) in order to theorize how contemporary print field guides enlist bird-watchers in Wittgensteinian language games. Lynch and Law argue that the use of field guides to create satisfying, name-producing encounters with birds simultaneously sustains and frustrates acts of visual field identification. Though Lynch and Law show that identifying all species by sight is sometimes impossible (Cooper's and sharp-shinned hawks, their example, can be particularly indistinguishable), the logic of the field guide perpetually functions to erase such problematics in support of the genre's authority. Taxonomic discourse is what conditions field guides so singularly toward the simple and problematic species classifications that Lynch and Law describe.

As taxonomic discourse, image and text in contemporary field guides actively engage in the systematic classification of organisms. Taxonomic discourse has a long history, most closely associated with systematic Linnaean classification in the eighteenth century. As Mary Louise Pratt describes (1992) and art from encounters with colonial landscapes demonstrates (Maria Sibylla Merian's illustrations, Catesby's paintings of birds), the taxonomic classification of kingdom, phylum, class, order, family, genus, and species was an imperialist project aimed at quantifying unknown landscapes (including humans in those landscapes) prior to exploitation. Describing the later uptake of taxonomic discourse by imagistic and narrative forms of nineteenth-century natural history, Margaret Welch writes:

> The practice and language of natural history reinforced a way of ordering the world that focused on species, with the individual

encountered representing the whole. The characteristics of appearance, locality, and habits that distinguish it from like species receive primary attention. Passages often list physical identifiers such as size in numerical figures, include Latin, and, in the longer prose sections, adopt the third person narrative. (1988, 166)

Several of the first American field guides to birds were works of natural history (Emily Taylor's *Conversations with the Birds* [1850], C. W. Webber's *Wild Scenes and Song-Birds* [1855], John Burroughs' *Wake-Robin* [1871/1904], Florence Merriam's *Birds through an Opera Glass* [1889], and Mabel Osgood Wright's *Birdcraft* [1895], for instance), and basing themselves upon this tradition, later twentieth-century field guides shed the nineteenth-century discourses of aestheticism, sentimentalism, and anthropomorphism to focus on the singular impetus of the taxonomic description. Two histories of the formation of American ornithology as a scientific discipline, Mark Barrow's *A Passion for Birds: American Ornithology after Audubon* (1998) and John Battalio's *The Rhetoric of Science in the Evolution of American Ornithological Discourse* (1998), show how contemporary bird-watching archives the late-nineteenth-century American ornithological preoccupation with classification. While taxonomic descriptions had been only one aspect of natural history writing, listing distinguishing features became formulaic for species descriptions in quick-reference guides. Defining taxonomic discourse as discrete from other classificatory discourses is not without problems, however, as the taxonomic imagination (Ritvo 1997) has corollaries to other forms of classification and differentiation, including language itself.

Alan Gross, in his examination of how taxonomic discourse was employed in the designation of a new species of the Central American hummingbird, describes how both written and visual representations can, by nature of being taxonomic, oversimplify differences between natural creatures while confusing the foundations of evolutionary theory (1990, 33–53). Studying the designation of this "new" hummingbird, Gross shows how species designation requires a "rhetorical construction" that

> redescribes the evolutionary species by translating into rhetoric each aspect of our rational reconstruction: no aspect we have described as science remains without its rhetorical counterpart. By means of rhetorical reconstruction, evolutionary taxonomy is transformed into an interlocking set of persuasive structures. *Sub specie rhetoricae*, we do not discover, we create: plants and animals are brought to life, raised to membership in a taxo-

nomic group, and made to illustrate and generate evolutionary theory. (1990, 34)

Contemporary field guides to birds, though only quasitechnical and focused on simple features bird-watchers can plainly see and observe, are just such rhetorical constructions. For Gross, the process of designating, naming, and rendering a species taxonomically on the written, drawn, and painted page amounts to "tendentious simplification" (1990, 45). Gross describes how the successful definition of the newly identified hummingbird species requires "a full-color painting of two of the birds *in situ*: a dimorphic pair, the female perched at an angle that most clearly displays her distinctive tail and underparts" as "the potential species is not only described and depicted in terms of family resemblances; it is also, by comparison and contrast, carefully differentiated from closely allied species. In this process of establishing species status, the character is the minimum unit of observation" (36). Such differentiation, through taxonomic images that represent and highlight differences, is borrowed from classificatory biology and foundational to the logic of contemporary field guide representation. Birds are not portrayed haphazardly in field guides, or anthropomorphically or naturalistically but strategically and mechanistically in order to focus on distinguishing characteristics and to sustain the differentiating habits of mind that bird-watching entails.

Gross could be describing the images in *The Sibley Guide to Birds* (2000), the National Geographic Society's *Field Guide to the Birds of North America* (2002), or any other contemporary field guide when he characterizes how a painting of the Central American hummingbird features specimens "clearly posed: it presents these creatures in a manner designed to display not just any characters, but only those that best distinguish them from their fellows" (45). While some of the first field guides to North American birds included plates depicting birds in naturalistic (even anthropomorphic) poses, contemporary field guides present images of birds aimed almost exclusively at differentiating each species to the point of uniqueness. It is this taxonomic discourse in contemporary North American field guides to birds that establishes connections and inheritances between contemporary field guides, bird-watching, and imperialist practices. While bird-watching is not overtly imperialist in most instances, as a popular pastime it is a form of amateur science that recreates critical features of imperialist encounters aimed at the quantification of lands subject to conquest. Inventorying the natural landscape in the imperialist encounter amounted to one way scientific discourse and activity were implicated in and used for land grabbing and the extraction of natural resources; at the same time, inventorying the natural landscape in the socio-rhetorical logic of bird-watching engenders pleasure and maintains a sense that the activity of bird-watching has environmental preservationist import. Bird-watching reenacts imperialist encounters while reanimating related practices with new meanings.

The Sibley Guide to Birds

Turning now to Figures 4.1 and 4.2 from *The Sibley Guide to Birds* (2000), my assertion of the pervasiveness of taxonomic discourse in contemporary field guides can be demonstrated. Before referring to

Figure 4.1. Thayer's and Iceland gulls from *The Sibley Guide to Birds*

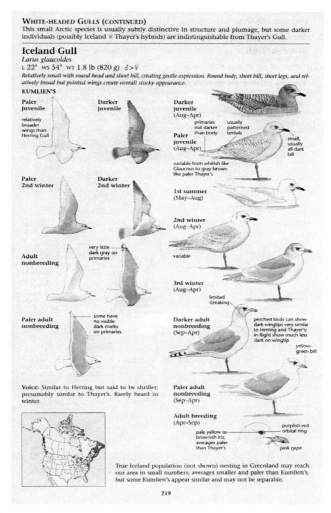

Figure 4.2. Iceland gull from *The Sibley Guide to Birds*

David Allen Sibley's field guide as typical of contemporary field guides, though, I should note some of the marked distinctions it has from other exponents of the genre. *The Sibley Guide* is physically larger than typical pocket-size guides, although Sibley has published (in 2003) smaller guides to the birds of North America divided into Eastern and Western regions. Furthermore, though similar to the majority of contemporary field guides in that Sibley's large guide uses paintings and not photographs, Sibley paints using watercolors, invoking a slightly more impressionistic tone to his images.[1] Even more importantly, the layout and design of text and image in *The Sibley Guide* depart from all other field guides since Peterson's popular guide of 1934, a genealogy of production that is well described in Raymond Korpi's dissertation *"A Most Engaging Game": The Evolution of Bird Field Guides and Their Effects on Environmentalism, Ornithology, and Birding, 1830–1998*. Juxtaposing Roger Tory Peterson's naturalistic field guide images with the more subversive, environmentalist images of Jack L. Griggs in his guide *All the Birds*, Korpi shows how environmental discourse has largely diminished in contemporary field guides. This diminishment is lamented by Korpi and seen as a depoliticization of the pastime. Whereas several contemporary field guides present image and text as separate units either on facing pages or on the same page (text at left and image at right, for instance, or top of the page for image, bottom for text), *The Sibley Guide* integrates image and text within vertical columns. Gunther Kress refers to this type of document design, of which *The Sibley Guide* is but one example, as "a new code of writing-and-image, in which information is carried differentially by the two modes" (1999, 76) working together to express a larger whole. Kress goes on to argue that the integration of image and text in print media signals that "boundaries of the 'text' are dissolving, and reading and use [of a text for specific purposes] become both blurred and fused" (77). In Sibley's large, comprehensive field guide, image and text certainly work together in the ways Kress describes, eschewing separate acts of reading text and image. Further facilitating this commingling of otherwise separate image and text sections is the fact that both elements are singularly taxonomic in import.

Figures 4.1 and 4.2 feature a page set (Figure 4.1 is verso in the set, Figure 4.2 recto), including written descriptions and descriptive images of two "closely related" species, the Thayer's and Iceland gulls. Sibley's description of white-headed gulls, at the top of page 218 (Figure 4.1), states that the two species "are often considered a single species that ranges from the slightly longer, darker Thayer's in the west to the smaller, paler Iceland in the east." With this admission, Sibley concedes that species classification is not as fixed and agreed upon as discrete naming signifies. I call this a concession within the context of Lynch's and Law's argument, mentioned earlier, that field guides overly distinguish indistinguishable species to promote their own authority and the viability of visual identification. Given this caveat, Sibley's text goes on to describe the observable differences that can be used to differentiate the two species from other gulls and each other. In this way—with a keen attention to noticeable difference—the discourse of the written descriptions is taxonomic. Nearly all text is dedicated to the quantification of features that distinguish, define, and classify. The Thayer's gull, for instance, has a voice that is "Lower and flatter than herring [gull]. Long call muffled, lower than herring gull; notes simple and flat, monotone. Rarely heard in winter." Other features of taxonomic distinction include the Thayer's gulls having an "often darker gray mantle than Iceland" in the adult nonbreeding population. Further, "Perched [Iceland gulls] can show dark wingtips very similar to Herring and Thayer's; in flight show much less dark on wingtip." While Sibley does not present complete taxonomies of either species, as would be the case in a field manual or ornithological text, the printed text on these pages is typical of contemporary field guide discourse in that it describes those taxonomic features that can be used by bird-watchers to distinguish, identify, and name birds in the field. Field marks of these kind are promoted to a position of highest importance by this discourse; features that are nondistinguishing have little value in a disposition that values successful identifications, even of gulls (*laridae*), an often overlooked family of birds.

The expressive conventions that inform Sibley's painted images are equally taxonomic, making for images that convey field marks. Visually, field marks are used to inform what Eugene Hunn has referred to, in his research on how bird-watchers identify gulls, as information enabling a "gestalt ID." That Sibley's images and others like them are taxonomic is an important point and essential to reading contemporary field guide images: the paintings in *The Sibley Guide* and other contemporary guides in circulation are not meant to thoroughly represent all features of each bird. For instance, both Thayer's and Iceland gulls have feet with complex and highly patterned textures, feathers with detailed patterns and shapes, heads that cast certain profiles when viewed at different angles, and many other characteristics that are of little use to distinction-minded bird-watchers. Features that fail to distinguish are either left out or minimized, as they offer little utility within the taxonomic imagination. Because some features are difficult to observe and of little classificatory significance, they are elided. Features that do "matter" when it comes to identification gain prominence in the visual representations. Such features as the coloration of the primaries, secondaries, beak, and mantle are presented clearly in the images of the Thayer's Gull; these are "what counts" within the classificatory logic of contemporary field guide discourse. Likewise, the bill, primaries, and "pale yellow to brownish" eye of the Iceland Gull (Figure 4.2) are carefully presented in the paintings of that bird, even including a close-up image that disembodies the trait.

Note that to hone in on such features and render the species visibly different and comparable, Sibley's birds almost all face right, pose in perfect profile, and stand, sit, or fly before vacant backgrounds. Rick

Wright decries the convention of painting very dead specimens as life-like, claiming that such practices of representation, present in all but one contemporary field guide (Ken Simpson and Nicolas Day's 1996 *The Princeton Field Guide to the Birds of Australia*), maintain the inaccurate belief that bird-watching is harmless to birds and fully divested from ornithology. As Sibley's guide demonstrates, contemporary field guides do not express the visual discourse of naturalism that locates birds in their "natural" landscapes, or the visual discourse of anthropomorphism that poses birds in what appear to be human predicaments and scenarios. Anthropomorphic images were common in late-nineteenth-century field guides, and naturalism still informs some contemporary guides today (the National Geographic Society's *Field Guide to the Birds of North America*, for instance, or Jack L. Griggs's and the American Bird Conservancy's *All the Birds*), but not the Sibley guide. Instead, a taxonomic discourse predominates, with Sibley aiming to present distinguishing features as quick to conceptualize, see, learn, and spot in the field. What these two species of gulls do on a beach, lakeside, or landfill is irrelevant to making a positive identification and thus left out of the textual descriptions and visual representations. What is presented is a maximum amount of diagnostic, taxonomic material in a finite amount of space.

Taxonomic Discourse Positioned as Reference

It is obvious to point out that different images can and usually do mean different things, and it is equally obvious to say that the same image in different contexts usually does not mean the same thing. The latter observation takes note of the relationship between semiosis and context. What I want to add here is a treatment of context as socio-rhetorical. Said another way, the socialities that different classes of reference materials are part of can have dramatic consequences on how their content is tendered. Textual positioning as I am thinking of it is concomitant with, but not necessarily equal to, genre. Genres locate, structure, organize, and encourage certain typified rhetorical actions (Miller 1984/1994); the positioning of a text (as art, or as a book of photographs, or as reference material) becomes a condition of the genre. Examples will help explain this.

Some artistic representations of birds include aspects of precision and attention to distinguishing field marks that are similar to images found in field guides. Take, for instance, the speculative paintings of Alexis Rockman. Rockman's paintings, displayed in a broad array of galleries and public spaces and reprinted in art and coffee-table books, express a vivid commentary on how humans have made and, Rockman predicts, will continue to make devastating changes to the planet. As art, this set of primary messages in Rockman's work is open to interpretation, circumspection, and ultimately levels of acceptance or rejection. In Rockman's paintings of birds (and birds figure prominently in Rockman's work; see "A Recent History of the World" for instance), the messages that "this is necessarily true" and "this is authoritative" are not immediately conveyed. Instead, the visual discourse is perspectival, subjective, and assumed to be personally expressive of the artist's primary concerns, interests, or vision. If there can be said to be a version of ecosee that Rockman's images about birds engender, then an adoption and a social circulation of this ecosee is preceded by consideration of whether or not the perspective presented in Rockman's paintings is one with which the viewer agrees and opts to embrace.

Consider the herring gull in Alexis Rockman's painting "Concrete Jungle V" (2003, 193), for instance. Even though David Allen Sibley's painting of this species includes similar distinguishing marks, the field guide image is imbued with a different character of authority given its positioning as a reference work located within the socio-rhetorical activity of bird-watching. Reference materials, of course, are the texts to which we turn to learn established, codified facts. When content is located within a reference text, it becomes associated with a certain type of doxa (Bourdieu 1990). Though not ubiquitously, many reference materials are constructed and designed as a quick reference, making the speed of the textual encounter yet another factor that hinders circumspection of content. It is not merely the nature of the images in field guides, then, that contributes to the version of ecosee we find in bird-watching but the positioning of those images in such a way that gives them a vested authority. This vested authority renders field guides in some ways even more regulatory amid the social practice of bird-watching than other visual texts that bird-watchers use. Memoirs about bird-watching (Kenn Kaufman's 1997 *Kingbird Highway: The Story of a Natural Obsession That Got a Little Out of Hand,* for instance) certainly influence bird-watchers but not as foundationally as reference works that are positioned as essential in the way field guides are. Field guide content—how to identify birds and what identifying traits consist of—is taken for granted and naturalized within bird-watching's particular socio-rhetorical condition. Thus a larger impetus is promoted and sustained—that satisfying encounters with birds need be classificatory to some degree.

Unlike Rockman's art, the way that field guide imagery is positioned as and used for reference locates the primary messages of taxonomic field guide representations as inherently genuine, truthful, and authoritative.

Because art in field guides is not first considered art but artistically produced reference matter made to precisely represent nature itself, Sibley's visual discourse immediately but indirectly puts into circulation the underlying assumption that all birds, or nearly all birds, can and should be identified by sight. Birds, in this discourse, are *for* identifying; bird-watching is all about identification.

In addition to being positioned as reference materials and thus imbued with an assumption that they are "true" and/or highly accurate, Sibley's images and their counterparts in other contemporary field guides to birds are at the center of a pastime with a very high reference orientation, by which I mean that successful everyday enactments of the pastime require regular consultation of bird-watching reference material, whether in print, online, or in the form of portable electronic devices. This high-reference orientation is one shared by other hobbies such as geocaching (which relies on global positioning systems and maps), trainspotting (mediated by identifying charts, manuals, and photos), and mushrooming (with its own field guides).[2] Of course, not all interactions with a cache, train, mushroom, or bird are always reference dependent, as usually only interactions with the unknown and unclassified turn participants to reference work. Still, bird-watching is among a subset of hobbies relying heavily upon the successful use of reference materials in order to create leisure. This is, of course, active leisure and productive leisure (Gelber) but pleasurable as leisure nonetheless. This reliance on printed texts positions bird-watching field guides as particularly influential on the creation and ongoing maintenance of classificatory bird-watching dispositions.

Gunther Kress describes how highly visual, multimodal texts relegate different forms of expressive content, commingling printed words and visual images to create complex pages where the emphasis is "less on reading than on doing" (1999, 68). Information-rich, multimodal reference materials of this kind comprise texts to which readers turn to find something useful, glean a fact or detail, and read strategically for certain elements while discarding others. As with the image-rich textbooks that Kress analyzes, field guides are books that are used more than "read" in any traditional sense (as in from cover to cover), making for yet another way that field guide images gain influence in the socio-rhetorical matrix of bird-watching. Contemporary field guides, designed with a quick reference in mind, present users with readily digestible bits of visual and textual content, and bird-watchers put to quick use these consumable bits in the classification and naming of unknown species of birds. The speed of the interaction, and the fact that field guides commingle image and text in the composition of books that "do" something (to borrow language from Kress), means that the taxonomic insistence of field guides has even more force.

Bird-watching in the ways that are suggested (and even sanctioned) by field guides means considering birds nature particles removed from complex ecosystems and needing to be identified without exception. Clearly the use of high-powered binoculars and spotting scopes aids in this type of particularizing vision, but field guides work with high-powered optics in complementarity, the one requiring the other. Identification is not all that bird-watchers do, of course, but it is the necessary first step they go through in order to "have" a named bird that can be discussed, appreciated, or noted on a list. If the first step of identification is not accomplished, then bird-watchers report frustration ensues (Kaufman 1997; Cocker 2001; Cashwell 2003). The nature of field guides as reference materials, the high reference orientation of the pastime, and the contributing factors of textual design that make for a quick reference and use of the guides facilitate the cultural uptake and circulation of taxonomic discourse in the form of thinking, speaking, listing, and acting. The images in field guides have so been naturalized as figments of the real that looking for field marks when looking at birds has become a commonplace way of seeing.

Some Conclusions

Ecosee functions broadly and variably across texts and scenes of interaction. The case of bird-watching and the genre of field guides demonstrate how that variation is contextual, with context here understood as involving how participants make meaning with texts. The intricacies of bird-watching as an established pastime with traditions and patterned forms of behavior inform how field guides are considered, used, and taken up as source material for this version of ecosee. In bird-watching, images and visual texts are positioned with authority at the center of environmental activity, making claims to influence more credible. In the pages of contemporary field guides to birds, image and text combine to form a visual apparatus to which users refer, flip through, and use in the completion of work specified by the pastime: accurate identification of birds in the field. As I have shown, using the example of *The Sibley Guide to Birds*, taxonomic features predominate in the visual landscape of contemporary field guides, propagating taxonomic lenses for looking at and distinguishing birds. These lenses, to maintain the visual metaphor, were once integral to imperialist practices of inventorying "unknown" landscapes to facilitate ecological exploration and exploitation.

Seeing bird identification as a reenactment of that history is masked, however, by the ways bird-watching encounters are currently motivated by aesthetic appreciation and a sense of doing something "environmental." Taxonomic discourse is asserted upon bird-watchers through the socio-rhetorical positioning of quick-reference texts as indispensable and central in textually mediated environmental encounters with birds. But what about those bird-watchers who work with but also around the influence of taxonomic discourse and classification? What can be said about bird-watchers who operate with a taxonomic imagination but do not make a project out of compulsively identifying every bird they see? This brings me back to the example of gulls.

I have argued that textually mediated forms of bird-watching relinquish themselves to regulated forms of looking that are largely taxonomic. This preoccupation with classification does not lend itself to bird-watching as a radical environmental intervention, however. As Raymond Korpi's (1999) work shows, it is only the rare field guide (Griggs's *All the Birds*) that portrays habitat destruction and the influence bird-watchers have on the landscapes they survey. Regulated, taxonomic forms of looking also are intrinsically moderate, because by rendering separate and distinguishable every species of bird, this form of ecosee enables and supports discourses of preference and "dis-preference." Though all contemporary field guides to birds devote relatively equal space on the page to all species, the separating and distinguishing impetus of the taxonomic imagination means that "doing gulls" or "not doing gulls" (by which bird-watchers mean working to classify members of that family of species or ignoring it) is enabled.[3] The distinguishing disposition that field guides promote cuts three ways: bird-watchers separate out birds from their larger landscapes, bird-watchers distinguish one species of bird from another, and some bird-watchers distinguish "trash birds" from more valued species. For many—and this is true for devotees and nondevotees of the pastime—gulls are a devalued class, and this "dis-preference" is informed by a number of discourses. One of these discourses is the taxonomic, as gulls pose exceeding difficulty for those habituated to the value systems and practices of bird-watching. Gulls, like flycatchers, sparrows, and a few other groupings of species, challenge efforts to classify using the field mark system, and in challenging this central aspect of bird-watching they create a backlash against gulls as a whole. The disparagement that gulls receive by some bird-watchers is in part sustained by the same discourses informing a classificatory consciousness. In this way, the highly visual and textually mediated disposition of bird-watching quivers with its imperialistic inheritances, as elements of the landscape that have been separated out from one another are evaluated once again. This time the evaluation does not aim to assess monetary value for a new empire; instead it participates in the ongoing maintenance of a classifying disposition.

Notes

1. Three popular contemporary guides that use photographs are the *National Audubon Society's Field Guide to North American Birds*, the *Stokes Field Guide to Birds*, and Kenn Kaufman's *Birds of North America*.

2. For more on the formation of hobbies in the United States, see Gelber 1999.

3. Though bird-watchers will note, the bald eagle nearly always receives a page of its own, at least since the reputation of that species was ameliorated in recent decades.

Works Cited

Barrow, Mark V. 1998. *A Passion for Birds: American Ornithology after Audubon.* Princeton, NJ: Princeton University Press.

Battalio, John T. 1998. *The Rhetoric of Science in the Evolution of American Ornithological Discourse.* Stamford, CT: Ablex.

Blanchan, Neltje. 1887/1897/1932. *The Bird Book: Bird Neighbors.* New York: Literary Guild of America.

Bourdieu, Pierre. 1984. *Distinction: A Social Critique of the Judgment of Taste.* Cambridge, MA: Harvard University Press.

———. 1990. *The Logic of Practice.* Stanford, CA: Stanford University Press.

Burroughs, John. 1871/1904 *The Writings of John Burroughs, Wake Robin.* Autograph ed. Vol. 1. Boston and New York: Houghton, Mifflin. Cambridge, MA: The Riverside Press.

Cashwell, Peter. 2003. *The Verb "To Bird": Sightings of an Avid Birder.* 1st Paul Dry Books ed. Philadelphia, PA: Paul Dry Books.

Catesby, Mark. 1722–1725. *The Natural History of Carolina, Florida, and the Bahama Islands.* Printed by expense of the author, and sold by W. Innys and R. Manby. London.

Cocker, Mark. 2001. *Birders: Tales of a Tribe.* New York: Grove Press.

Gelber, Steven M. 1999. *Hobbies: Leisure and the Culture of Work in America.* New York: Columbia University Press.

Gleason, William A. 1999. *The Leisure Ethic: Work and Play in American Literature, 1840–1940.* Stanford, CA: Stanford University Press.

Griggs, Jack L., and American Bird Conservancy. 1997. *All the Birds of North America: American Bird Conservancy's Field Guide.* 1st ed. New York, NY: Harper Perennial.

Gross, Alan G. 1990. *The Rhetoric of Science*. Cambridge, MA: Harvard University Press.

Hunn, Eugene. 1975. "Cognitive Processes in Folk Ornithology: The Identification of Gulls." *Working Papers of the Language Behavior Research Laboratory, University of California, Berkeley*. Language Behavior Research Laboratory, University of California. Berkley, CA. Paper no. 45.

Karnicky, Jeffrey. 2004. "What Is the Red Knot Worth?: Valuing Human/Avian Interaction." *Society & Animals Journal of Human-Animal Studies* 12.3 (2004), retrieved August 2005 from http://www.psyeta.org/sa/sa12.3/karnicky.shtml.

Kaufman, Kenn. 1997. *Kingbird Highway: The Story of a Natural Obsession That Got a Little Out of Hand*. Boston, MA and New York: Houghton Mifflin.

Korpi, Raymond Thomas. 1999. *"A Most Engaging Game": The Evolution of Bird Field Guides and Their Effects on Environmentalism, Ornithology, and Birding, 1830–1998* Dissertation. Washington State University, American Studies Program.

Kress, Gunther. 1999. " 'English' at the Crossroads: Rethinking Curricula of Communication in the Context of the Turn to the Visual." In *Passions, Pedagogies, and 21st Century Technologies*, ed. Gail E. Hawisher and Cynthia L. Selfe, 66–88. Logan: Utah State University Press.

Lynch, Michael, and John Law. 1999. "Pictures, Texts, and Objects: The Literary Language Game of Bird-Watching." In *The Science Studies Reader*, ed. Maria Biagioli, 317–41. New York and London: Routledge.

Merriam, Florence A. 1889. *Birds through an Opera-Glass*. New York: The Chataqua Press (Houghton Mifflin).

Miller, Carolyn R. 1984/1994. "Genre as Social Action." In *Genre and the New Rhetoric*, ed. Aviva Freedman and Peter Medway, 23–42. Critical Perspectives in Literacy and Education. London: Taylor and Francis.

National Geographic Society (U.S.). 2002. *Field Guide to the Birds of North America*. 4th ed. Washington, DC: National Geographic.

Peterson, Roger Tory. 1934. *A Field Guide to the Birds*. 1st ed. Boston, MA, and New York: Houghton Mifflin.

Pratt, Mary Louise. 1992. *Imperial Eyes: Travel Writing and Transculturation*. London and New York: Routledge.

Ritvo, Harriet. 1997. *The Platypus and the Mermaid and Other Figments of the Classifying Imagination*. Cambridge, MA: Harvard University Press.

Rockman, Alexis, et al. 2003. *Alexis Rockman*. New York: Monacelli Press.

Sibley, David Allen. 2000. *The Sibley Guide to Birds*. New York: Chanticker Press, Knopf.

Simpson, Ken, and Nicolas Day. 1996. *The Princeton Field Guide to the Birds of Australia*. 5th ed. Princeton, NJ: Princeton University Press.

Taylor, Emily. 1850. *Conversations with the Birds*. Salem: W & S.B. Ives.

Webber, C. W. 1855. *Wild Scenes and Song-Birds*. New York: Riker, Thorne, and Co.

Welch, Margaret. 1998. *The Book of Nature: Natural History in the United States, 1825–1875.* Boston, MA: Northeastern University Press.

Wright, Mabel Osgood. 1895. *Birdcraft: A Field Book of Two Hundred Song, Game, and Water Birds.* New York: Macmillan and Co.

5

Eduardo Kac

Networks as Medium and Trope

Simone Osthoff

In the future much more than the simple defense of nature will
be required.

—Félix Guattari, *The Three Ecologies*

Over two decades, Eduardo Kac's hybrid networks have connected in real
time disparate and distant elements. They also have offered new insights
into art, while leading the artist in 1999 to the literal creation of new
hybrid life-forms. By changing habitual ways of seeing and communicating, Kac's networks and transgenic creations continuously challenge our
understanding of the "natural" environment as well as the environment
of art. They explore what French philosopher Jacques Rancière called
the "distribution of the visible, the sayable, and the possible."[1] This
chapter offers a brief overview of Kac's development from the early
1980s, focusing on two telepresence works of the mid-1990s—*Rara
Avis* and *Time Capsule*. A third focus is the juxtaposition of the artist's
2004 *Rabbit Remix* exhibition and the publication of two anthologies
of the artist's writings. In my conclusion, I argue that Kac's theoretical
essays constitute an intrinsic part of his networked ecology.

By converging art, science, and technology with communication
theory, philosophy, and poetry, the artist produces unusual connections
such as those among language, light, and life. Insightful and experimental, Kac's work suggests alternative ecologies neither by denouncing
climate change and environmental disasters nor by calling attention to
monstrous threats produced by the manipulation of DNA information.
The dimensionalities and temporalities explored by Kac's networks—both

human and nonhuman—examine the wider ecological questions posed by Félix Guattari's *The Three Ecologies*, a manifesto that called upon activists "to target the modes of production of subjectivity, that is, of knowledge, culture, sensibility, and sociability."[2]

Prompting a continuum between nature and culture, between species, and among the senses, Kac's work questions the structures, mediations, and ultimately the supremacy of vision in art, while promoting synesthetic experiences that rearticulate individual consciousness within social, cultural, and finally environmental realms. In addition, his work addresses issues of spectatorship by emphasizing participatory action and two-way communication. Kac's hybrid networks of physical and virtual spaces dislocate audiences within environments that examine how vision, touch, hearing, and voice are facilitated and constrained by the structures and mediation of technology. Within his networked environments, dialogical communication among humans, animals, plants, microorganisms, and machines is never given but instead must be construed by participants word by word, frame by frame.

Kac's Twenty-Five-Year Trajectory: Connecting Language, Light, and Life

Never purely visual, always impurely polysemic, and disregarding traditional disciplinary boundaries, the artist's works are neither easy to classify nor to locate. When I first interviewed Kac more than a decade ago, being curious about his fluency in at least four languages (English, Portuguese, Spanish, and French), I began by asking him about his nationality. He answered that his work was not about location but connectivity: "I prefer not to be bound by any particular nationality or geography. I work with telecommunications, trying to break up these boundaries."[3] For him, identity and location are never fixed but vectors in the production of subjectivity that his work explores.

Kac began his career in Rio de Janeiro with wildly transgressive poetry performances on Ipanema Beach (1980–1982). In 1983, seeking to create a new language for poetry out of the fluidity of light, the artist found in holography a new medium for art making. His holopoems (1983–1993), which use light as an immaterial writing environment, depend on the location of the body of the viewer in space for the construction of their syntax and semantic meanings. Kac approached holography as a time-based medium, where both the eyes and the whole body of the viewer are activated.

Parallel to his holopoems, since 1985 Kac has been exploring communication at a distance in complex interactive works connected via telecommunication systems—at first videotexts, videophones, and telerobotics—then through more complex networked events taking place on the Internet. In these telepresence works, communication was not only mediated by hardware and software but was negotiated among multiple participants, not always human, often of different species, such as in his 1994 *Essay Concerning Human Understanding*, in which a dialogue took place through the network between a plant in New York and a bird in a gallery in Lexington, Kentucky. Many viewers were skeptical. Was the work a practical joke? Could a real phone dialogue take place between a plant and a bird? Was this a poetic metaphor or a real and literal conversation?

Actual communication involving not only different species but also multiple institutional nodes both private and public is central to the artist's aesthetics. Kac's telepresence events emphasize real time over real space, linking humans, animals, plants, and machines in several nodes of observation and participation worldwide. Furthermore, his telepresence events underline the spatial dislocation of vision into multiple points of view. Between Kac's telepresence events with the *Ornitorrinco* telerobot (1989–1996) and the transgenic creations he started in 1999 are a number of complex telepresence installations and performances that expanded the artist's examination of interspecies and remote communication also including human-machine exchanges, such as: *Teleporting an Unknown State* (1994–1996); *The Telepresence Garment* (1995–1996); *Rara Avis* (1996); *A-Positive* (1997); *Time Capsule* (1997); *Uirapuru* (1996–1999); and *Darker Than Night* (1999).

Commenting on a series of telepresence works created with his telerobot *Ornitorrinco*, Kac stated: "What the telepresence installation with the Ornitorrinco telerobot is all about is to metaphorically ask the viewer to look at the world from someone else's point of view. It's a non-metaphysical out-of-body experience, if you will."[4] This positional exchange between viewers and the point of view of the telerobot is further expanded in *Rara Avis*, 1996, where the artist employed VR technology and multiple Internet protocols to displace the viewer's gaze into the body of a robotic macaw while turning that gaze upon viewers themselves.

A transitional work between Kac's telepresence events and his bio art is *Time Capsule*, from 1997, in which the artist examined issues of memory and digital archives, that literally entered the artist's body through the implantation of a microchip in his ankle. The chip information was then interactively stored in a data bank in the United States while being simultaneously broadcast on television and on the Web. Like his early performances on Ipanema Beach, holopoems, and telepresence events, Kac's bio artworks continue to explore communication processes, as well as new ways of seeing, writing, reading, and speaking. Translation and inscription are especially prominent in his transgenic Creation Trilogy—*Genesis* (1999), *GFP Bunny* (2000), and *The Eighth Day* (2001)—and in the also trangenic *Move 36* (2002/2004).[5] However, since 2004, in his *Rabbit Remix* ongoing series, Kac employs the media reception and circulation of his work across space and time as a new material for art making, thus re-defying and enlarging the concept of network.

Network as Medium: The Examples of *Rara Avis* and *Time Capsule*

As part of his dialogic practice, Kac often forged new venues for his work and thus approached art institutions less as a hardware and container of culture and more as interface, where institutions might function as software, frame, or site—one more node of his ephemeral ecologies. For instance, when in 1996 Kac was invited by Nexus Contemporary Art in Atlanta (now Atlanta Contemporary Art Center) as part of the cultural events surrounding the Olympic Games, that important and large art center only had telephone and fax machines. Kac brought the Internet to their galleries for the creation of the *Rara Avis* installation, which networked the Nexus Contemporary Art to the Internet through three protocols: CU-SeeMe, the Web, and the MBone. In this work, local and remote participants experienced a large aviary from the point of view of a telerobotic macaw placed among thirty live birds.[6]

The VR technology *Rara Avis* employed was state of the art in 1996, and Kac used it to subvert common expectations about immersive technologies. Instead of offering a simulation, he turned the viewers' gaze back upon themselves by projecting in real-time stereoscopic 3-D color images of the viewers in the gallery. The two cameras were located in the eyes of the exotic robotic bird within the aviary. Gallery visitors wearing the VR headset saw from close up the thirty live birds flying, eating, and perched on branches quite close to the camera lenses. They simultaneously saw themselves in the image background, standing outside of the aviary wearing the VR headset, and thus being both inside and outside of the aviary at the same time. Kac summed up this reversal: "In *Rara Avis*, the spectacular became specular, forcing the viewer to see himself or herself through the eye of the so-called exotic being."[7]

While denying users of VR technology the simulated worlds they normally expected and instead offering them a reflective visual experience of simultaneously seeing and being seen, the artist in addition networked the headset images with remote participants on the Web. Participant-

Figure 5.1. Eduardo Kac, *Rara Avis*, 1996. Telerobot, thirty zebra finches, aviary, VR headset, Internet

viewers elsewhere saw on their computer screens the same real-time video images projected inside the VR headset, and if they used their own home cameras and the CU-SeeMe program, they also could send live audiovisuals of themselves, thus seeing and talking to other users. Strange conversations took place. As a privileged viewer-participant of *Rara Avis* in three of its exhibition venues (I was present in the gallery in Atlanta in 1996, at the Mercosul Biennial in Porto Alegre, Brazil, in 1997, and as a remote viewer-participant using CU-SeeMe during the opening of *Rara Avis* in Austin, Texas), I experienced firsthand not only the uncanny dislocation of point of view the work promoted for viewers in the gallery but also how the Internet stratification and fluctuating traffic patterns produced alternative experiences of the work. During the opening of the exhibition in Texas, sounds and images were being exchanged in real time among remote participants and combined with the voices and images of the viewers and birds in the Austin gallery. My remote reception of the sounds was not always synchronized with that of the images. Fluctuations and delays produced fragmented moving images and disjointed real-time conversations, which at some point included the feedback of my own voice as a ghostly presence in the gallery opening night.

Time Capsule was a more dramatic and controversial, yet equally bold, network installation and performance. After being censored by the first venue where the event was scheduled to take place—the Itaú Cultural in São Paulo—another prestigious venue in the city merely one block away from the first—the cultural center Casa das Rosas—offered to showcase the performance and exhibition.[8] The work consisted of a microchip implant, a live television broadcase, and a simultaneous Web cast of the performance, an interactive telerobotic Web scanning of the implant, a remote database registration, and additional display elements, including seven sepia-toned photographs and X-rays of the artist's ankle before and after the implant.

The performance-media-spectacle took place on November 11, 1997, when the artist implanted a memory chip in his own ankle in a gallery exhibition that displayed old sepia family photographs on the wall (the only images that surivived after his family had to flee Poland in 1939), thus bringing to the critical forefront of his work questions of information, documentation, and history, which have always been connected in Kac's artistic practice from the beginning of his career. Many journalists with cameras of all sizes filled the gallery documenting the microchip bio-implant simultaneously in the print media, in a live broadcast on TV, through later TV broadcast updates, and on the

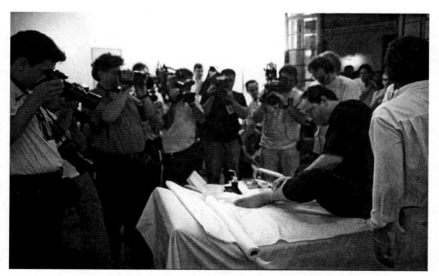

Figure 5.2. Eduardo Kac, *Time Capsule*, 1997, Microchip implant, simulcast on TV and the Web, remote scanning of the microchip through the Internet, photographs, X-ray

Web. After the insertion of the microchip in the artist's body, the digital information it contained was remotely retrieved by a scanner attached to a computer, as the artist registered the chip's ID number over the Internet in a databank in the United States.

Time Capsule locates a digital archive—a computer memory unit used to track animals—inside of the artist's body underlying the increasing embodiment of technology: "Scanning of the implant remotely via the Web revealed how the connective tissue of the global digital network renders obsolete the skin as a protective boundary demarcating the limits of the body."[9] In addition, *Time Capsule* occupied many sites simultaneously: the artist's body, the gallery space, the mass media, and the Web, as this memory archive traversed the skin boundary and thus blurred its inside and outside limits.

In a poignant keynote lecture philosopher Jacques Rancière connected the concepts of spectators, spectacle, theater, and intellectual emancipation. Rancière's words echo Kac's emphasis on a dialogic aesthetics: "Emancipation starts from the principle of equality. It begins when we dismiss the opposition between looking and acting and understand that the distribution of the visible itself is part of the configuration of domination and subjection. It starts when we realize that looking is also an action that confirms or modifies that distribution, and that 'interpreting the world' is already a means of transforming it."[10] By creating experiences where the points of view of observers are dislocated, distributed and unstable, viewers' awareness of their own position in the work may change in the passage from the experience of being an observer to that of becoming observed (by the birds, by Internet participants, by himself or herself through the eyes of the macaw). This complex examination of the act of seeing—which *Rara Avis* and *Time Capsule* promote and distribute among multiple sites—the exhibition *Rabbit Remix* will further extend by orchestrating the global media response that the *GFP Bunny* generated between 2000 and 2004.

Rabbit Remix: The Media as Medium

The media, understood both as the plural of medium, and as the means of mass communication (such as newspapers, magazines, radio, and television), have been explored as a medium by the artist since the beginning of his career. But beyond a few revolutionary and enthusiastic moments such as the one led by the Russian Constructivists between 1917 and 1925 and by the Bauhaus artists, in general the relationship between the twentieth-century avant-gardes and the mass media remained controversial—from the Futurist radio to Orson Welles's famous adaptation of the sci-fi novel *War of the Worlds* in a radio broadcast on the day prior to Halloween in 1938; and from Pablo Picasso Cubist's collages of 1912 to Jackson Pollock's 1949 photo spread on the pages of *Life* magazine and Andy Warhol's profitable engagement with celebrity and commodity culture.

Throughout the 1970s and in many parts of the world, video art began to further close this gap as visual artists increasingly embraced video as an experimental, time-based medium. Nam June Paik's *Global Groove* 1973 anticipated the MTV aesthetics and the Martha Rosler 1975 deadpan performance of domesticity in *Semiotics of the Kitchen* contrasts sharply with the visual exuberance of Mariko Mori's high-tech nirvana-pop videos of the 1990s, as well as with Mathew Barney's ambitious epic *Cremaster Cycle*, completed in 2002. Over the last fifteen years, artists such as Orlan and Stelarc have in different ways been highly skilled in framing their notorious performances as public media spectacles. Others, such as Andrea Zittel, have collapsed the boundaries between art and design by creating their own brand of products. And because of the early media acclaim she received, Cindy Sherman has had to negotiate her media image as another dimension of her identity and numerous self-portraits throughout her career.

Kac's performance is uncommon among artists and theorists, because he is fluent in multiple languages and fields of knowledge, ultimately influencing the history of new media as well as participating in the theoretical discussion that his work generates. Besides being an accomplished researcher and writer, Kac has always articulated the experience of creative work with aesthetic theory. Among the few artists who can lucidly speak about aesthetic concepts in relation to other disciplines, such as science, technology, and poetry, his voice contributes to debunk the fantasy that studio work does not involve either theory or research, thus grounding his creations both in experiment and debate. Always minding the cognitive structures of communication processes, Kac's networks and writings continuously connect art and life, culture and nature, and art writing and art making. An example is the juxtaposition of the publication of Kac's 1980s' critical writings in his book *Luz & Letra* and his solo exhibition *Rabbit Remix*, which linked the media interventions in Rio de Janeiro by the artist over a period of twenty-five years.[11]

In September 2004, when I arrived in Rio de Janeiro on my way to the 26th São Paulo Biennial, images of Eduardo Kac's *GFP Bunny* were strategically placed throughout the city on three types of advertising displays: illuminated advertising signs mounted above digital clocks/thermometers put on view the enigmatic green bunny, panels at bus stops announced his solo exhibit at Laura Marsiaj Gallery in Ipanema, and constantly moving displays rotated images of cultural events in the city, among them Kac's *GFP Bunny* and Bebel Gilberto's new CD. One week later, at the São Paulo Biennial, Kac presented a transgenic installation titled *Move 36*,[12] which was being appointed by the media as one of the must-sees among the 135 artists from sixty-two countries of this mega-event. Interviews and images of his installation appeared in the major newspapers and magazines of Rio de Janeiro and São Paulo prior, during, and after the opening of the exhibit.[13]

Besides being the title of his solo show in 2004, *Rabbit Remix* also titles an ongoing series of works with three phases: the first was the creation of the *GFP Bunny* in 2000; the second was the *Free Alba!* campaign carried out by the artist in 2001–2002; and the third is his ongoing orchestration of the ensuing global media response to this work. The *Rabbit Remix* series extends the discussion of bio art in relation

Figure 5.3. Digital street clock in Ipanema Beach with image of Kac's 2000 *GFP Bunny*, a public intervention in Rio de Janeiro as part of his solo show *Rabbit Remix* at the gallery Laura Marsiaj Arte Contemporânea, Rio de Janeiro, Brazil, 2004 (artwork by Eduardo Kac; photograph by Nelson Pataro; provided by the artist)

to science, ethics, religion, and culture, which Kac continues to address beyond the space of the gallery in many forms such as mass-media articles and interviews, academic books and essays, lectures and debates, and public interventions.

"Information is never found in a pure state. It always implies a point of view," observed Kac in reference to his 2004 exhibit *Rabbit Remix*.[14] An important component of this exhibition at the Laura Marsiaj Gallery was its advertising campaign as a further intervention in the public space of Rio de Janeiro—the scene where the artist first started reclaiming the public space in the early 1980s. The gallery exhibition was comprised of a series of photographs, drawings, a flag, a Web piece, and a limited-edition artist's book titled *It's Not Easy Being Green!* (most of the large photographs are now in the Gilberto Chateaubriand Collection of the Museum of Modern Art, Rio de Janeiro). Kac's remix of the *GFP Bunny* icon, which includes the reappropriation of the media response to his work, both verbally and visually, employs *the media as a medium.*

The publication of Kac's two volumes of collected writings and essays, which stress Kac's performance and voice as an artist-theorist, coincided with his 2004 *Rabbit Remix* exhibition. The first compilation, published in Brazil also in 2004, is titled *Luz & Letra: Ensaios de Arte, Literatura e Comunicação* [Light & Letter: Essays in Art, Literature, and Communication], thus far in Portuguese only. It collects Kac's early articles and essays written between 1981 and 1988 and published in the most important newspapers in Rio de Janeiro and São Paulo, along with an appendix of early projects and sketches. The second anthology, *Telepresence and Bio Art: Networking Humans, Rabbits, and Robots* (2005), compiles articles published in the United States between 1992 and 2002. The subtitle of Kac's 2005 book—*Networking Humans, Rabbits, and Robots*—underlines the artist's radical and hybrid connectivity in which, I argue, the books themselves are constitutive elements.

The articles included in *Luz & Letra*, originally written to discuss and promote electronic art, had a lasting impact. In their visionary originality they are early critical probes at the intersection of art, literature, technology, and popular culture. Written in an elegant, direct, and informative style, Kac's articles and essays challenged established artistic notions, values, and venues. In the Preface of *Luz & Letra*, art critic Paulo Herkenhoff, a former curator at the Museum of Modern Art in New York, stresses the importance of Kac as a theoretician:

This book is a document which recovers the 1980 decade—a period thought to be lived under the tyranny of painting—as a moment of gestation of new ideas. *Luz & Letra* reveals a

mode of thinking about a contemporary practice in parallel to
the traditional processes of artistic production. It points to, in
sum, a double cultural state: the degree of discussion of media
art in Brazil and the capacity of an artist to critically absorb
and interpret the possibilities of technology. Eduardo Kac is a
precursor among precursors of media art theory [. . .]. To him
the central question, however, was never his own placement in
history. On the contrary, his action was always characterized by
an intention to alter a system of hierarchies through the rescuing
of artists and experiences, which encourage the construction of
new negotiating processes about the presence of art in society.
From this horizon, artists of interest emerge from their humble
and discrete status, from their silence, exclusion, or exile.[15]

The juxtaposition of the publication of *Luz & Letra* with Kac's
exhibition *Rabbit Remix* revealed a direct relationship from the beginning of his career among his work, his critical writings, the gallery
space, and the mass media. In September 2004, all of these multiple
arenas were occupied simultaneously by the glowing rabbit icon as it
continued its four-year rapid propagation along with a controversy of
unforeseen scale and speed.

Network as Trope: Interpolations of the Artist-Theorist

Like numerous conceptual and performance-based artworks, Kac's
networks further erase boundaries between the artwork and its documentation, thus frequently challenge the traditional separation between
the artist, the historian, and the critic. Yet unlike much art of the last
forty years, Kac is not interested in metaphorical images but in actual
experiences that include live and remote communication as well as live
hybrid beings (he has created new living organisms specifically for each
new artwork since 1999). Given the unprecedented nature of Kac's creations in concert with his clear and articulated writings, my suggestions
of his use of network as a trope may indeed deviate from the literal
sense the artist privileges in his art making. Nevertheless, the artist's
voice—which beyond his works is present in his theoretical writings,
lectures, and interviews—clearly constitutes further hybrid interpolations
in his networks, pointing out, in addition to the artworks themselves,
important philosophical, aesthetic, and critical questions.

While the best artists' writings explore what constitutes medium,
both conceptual and creative processes, and institutional context,

increasingly artists have had to fill the gaps left by art criticism.[16] If there is a common agreement in current discussions of art criticism it is the recognition of a general crisis, as foreground by the 2002 *October* group roundtable "The Present Conditions of Art Criticism," by James Elkins's 2003 booklet *What Happened to Art Criticism?*, by Raphael Rubinstein's 2003 article "A Quiet Crisis," and by Nancy Princenthal's 2006 article "Art Criticism, Bound to Fail."[17] As many agree, the expansion of the global art market has paradoxically diminished the ability and interest of critics to make value judgments. Therefore, art criticism has become increasingly more informative and promotional than critical, and art history, for the most part, continues the tradition of lagging behind new media artists' own theoretical articulations.

In the Foreword of Kac's *Telepresence and Bio Art*, art historian James Elkins pointed out: "This is an unusual book, because Kac has participated in the movements he discusses. He is an artist and also, at times, a historian. The combination is rare." Elkins is right in positioning Kac as a historian "at times," because most of the time the artist is a theoretician. In his writings the historical research is at the service of his theoretical argumentation.[18] Kac's 2005 book articulates several new concepts he introduced, such as *telepresence art, telempathy,* and *performative ethics.*

Formerly, artists' writings were included within the discourses of art history as source material, but not as authoritative voices. Nevertheless, they continue to disrupt these discourses from within. From Marcel Duchamp's *Fountain* to Kac's *GFP Bunny*, revolutionary artistic practice often exposes art's taboos, biases, and ideological frames. Kac's networks examine how technology-mediated environments structure our perception and cognition. By approaching networks as a medium and trope, that is, by displacing visual perception and the clear location of voice and vision through viewer participation and interpolation in his networks (in which I include the effects produced by the artist's own voice and discursive body of work), Kac places aesthetic experience at the center of philosophical concerns, as a few philosophers also have privileged, among them Kant, Adorno, and Rancière. Kac's networks change the hierarchies and the function of institutions and the clear location of voice and vision among network participants, ultimately including our cognitive understanding of the "natural" within the environment of art.

Notes

1. "Art of the Possible, Fulvia Carnevale and John Kelsey in conversation with Jacques Rancière, *Art Forum* (March 2007): 256–59.

2. Félix Guattari, *The Three Ecologies* (London: Athlone Press, 2000), 49.

3. Simone Osthoff, "Object Lessons," *World Art* (Spring 1996): 18.

4. Kac, *World Art*, 22.

5. All of Kac's transgenic works up to and including *Move 36* are documented in Eduardo Kac, *Telepresence & Bio Art: Networking Humans, Rabbits, & Robots* (Ann Arbor: University of Michigan Press, 2005). About *Move 36*, see also Elena Rossi, ed., *Eduardo Kac: Move 36* (Paris: Filigranes Éditions, 2005), with essays by David Rosenberg, Frank Popper, Didier Ottinger, Linda Weintraub, and Hugues Marchal.

6. *Rara Avis* premiered as part of the exhibition Out of Bounds: New Work by Eight Southeast Artists, curated by Annette Carlozzi and Julia Fenton (Atlanta: Nexus Contemporary Art Center, June 28–August 24, 1996). In 1997, *Rara Avis* traveled to three other venues: the Jack Blanton Museum of Art, Austin, Texas; the Centro Cultural de Belém, Lisbon, Portugal; and the Casa de Cultura Mario Quintana, Porto Alegre, Brazil, as part of the *I Bienal de Artes Visuais do Mercosul*.

7. Kac, *Telepresence & Bio Art*, 163.

8. Patricia Decia. "Bioarte: Eduardo Kac tem obra polêmica vetada no ICI," *Folha de São Paulo* (October 10, 1997), Ilustrada, 13.

9. Kac, *Telepresence & Bio Art*, 232.

10. Jacques Rancière, "The Emancipated Spectator," *Art Forum* (March 2007): 271–80.

11. Eduardo Kac, *Luz & Letra: Ensaios de Arte, Literatura e Comunicação* [Light & Letter: Essays in Art, Literature, and Communication] (Rio de Janeiro: Contra Capa, 2004).

12. Besides the São Paulo Biennial, *Move 36* also was exhibited in 2004 at the Gwangju biennale in Korea and in 2005 in Paris at the Biche de Bere Gallery, September 28–October 26.

13. Among many others, see Caroline Menezes, "Uma nova genética para a arte. Eduardo Kac usa genes para discutir relação entre ser vivo e tecnologia," *Jornal do Brasil* (September 30, 2004), Caderno B, B4; Giselle Beiguelman, "O xeque-mate cibernético," *Folha de São Paulo* (September 19, 2004), Caderno Mais!, 14–15; "A Coelha Transgênica," *Veja Rio* (September 22, 2004), 43.

14. Eduardo Kac's public lecture at the gallery Laura Marsiaj, Rio de Janeiro, September 20, 2004, on the occasion of his show *Rabbit Remix*.

15. Paulo Herkenhoff, Preface of Kac, *Luz & Letra*, 18. Translation mine.

16. The combination of art practice with critical research and writing is about to become even more common in the United States over the next decades because of the upcoming PhD degree in studio art. The question of what constitutes research in fine arts studio education and the role of academic writing in such pursuit is open for debate, while it also points to new connections to be explored among previously unrelated academic fields. See the multiple contributions to this discussion by prominent artists and historians in "Art Schools: A Group Crit," *Art in America*, (May 2007): 99–109. Further examples of accomplished artists who also are critical writers in the United States include

the following: Donald Judd, *Complete Writings 1959–1975* (Halifax, Nova Scotia: Press of the Nova Scotia College of Art and Design, 2005); Andrea Fraser, *Museum Highlights: The Writings of Andrea Fraser* (Cambridge, MA: MIT Press, 2005); Martha Rosler, *Decoys and Disruptions: Selected Writings, 1975–2001* (Cambridge, MA: MIT Press, 2004); Robert Smithson, *The Collected Writings*, ed. Jack Flam (Berkeley: University of California Press, 1996); Joseph Kosuth, *Art after Philosophy and After: Collected Writings, 1996–1990* (Cambridge, MA: MIT Press, 1991).

17. Rosalind Krauss, Benjamin Buchloh, George Baker, Andrea Fraser, David Joselit, Robert Storr, Hal Foster, John Miller, James Meyer, and Helen Molesworth, "Roundtable: The Present Conditions of Art Criticism," *October* 100 (Spring 2002): 200–28; James Elkins, *What Happened to Art Criticism?* (Chicago, IL: Prickly Paradigm Press, 2003); Raphael Rubinstein, "A Quiet Crisis," *Art in America* (March 2003): 39–45; Nancy Princenthal, "Art Criticism, Bound to Fail," *Art in America* (January 2006): 43–47.

18. Further editorial projects by Kac include: Eduardo Kac, ed., *Signs of Life: Bio Art and Beyond* (Cambridge, MA: MIT Press, 2007) and Eduardo Kac, ed., *Media Poetry: An International Anthology* (Bristol: Intellect, 2007), first published as a special issue of the journal *Visible Language* (1996) vol. 30, no. 2.

Part 2

Seeing Animals

6

From *Dead Meat* to Glow-in-the-Dark Bunnies

Seeing "the Animal Question" in Contemporary Art

Cary Wolfe

Introduction

This chapter begins at the intersection of two questions: one, apparently quite complicated; the other, apparently quite simple. The first question—which is invoked but not really articulated by the phrase "animal rights"—concerns the ethical standing of (at least some) nonhuman animals. It is a question with which we are confronted every day in the mass media—indeed, entire cable television networks are now built around the presumption of its possibility—and it has increasingly captivated not just scientific fields such as cognitive ethology, ecology, and cognitive science but also areas in the humanities such as philosophy, psychoanalysis, "theory," and cultural studies generally. For the purposes of this chapter, I am simply going to take it for granted that the ethical standing of at least some nonhuman animals is not just a live issue but an increasingly taken-for-granted one (even if how to formulate that ethical standing remains a complex question). And I am going to allow myself this luxury in no small part, because the two artists whose work I will be addressing take that standing for granted, as they have affirmed in a variety of contexts.

The second question seems, by comparison, much more straightforward, and perhaps almost trivial in comparison to the weight of the first, but that is part of the reason I want to take it up here: When contemporary artists take nonhuman animals as their subject—our treatment of them, how we relate to them, and so on—what difference does it make that those artists choose a particular representational strategy (and—a question I cannot fully explore here—a particular medium or art form such as painting, sculpture, installation, or performance, just to name a few). To put this a little more directly, there clearly has been in contemporary art an explosion of interest in what Derrida calls "the question of the animal" as theme and subject matter (Derrida 2002).[1] When addressing this topic, however, it is all too easy to fall into what Slavoj Zizek, with characteristic astringency, has in another context called "an un-dialectical obsession with content" (1994, 202). What I am interested in, on the other hand, is how particular artistic strategies *themselves* depend upon or resist a certain humanism that is quite independent of the manifest content of the artwork—the fact that it may be "about" nonhuman animals in some obvious way.

We can bring the question I have in mind into even sharper focus along the following lines: If, as many of the most important contemporary thinkers have suggested, certain representational strategies (say, the Renaissance theory of perspective, or Bentham's panoptical rendering of architectural space, or the production of the gaze and spectatorship in film as critiqued by feminist film theory in the 1980s, and so on) can be indexed to certain normative modes of humanist subjectivity that they reproduce *by the very nature of their strategies*, then we are well within our rights to ask—to put it succinctly, for the moment—what the relationship is between philosophical and artistic representationalism.

These are precisely the sorts of questions that practicing artists routinely engage in connection to the specific demands of particular representational media. And they bear very directly upon not just the artistic challenge but also the larger philosophical and ethical challenge, of speaking *for* nonhuman animals, speaking *to* our relations with them, and how taking those relations seriously unavoidably raises the question of who "we" are—a question that may get answered quite indirectly not in the manifest content of the artwork or its "message" but in its formal strategies.

The Ethics of (Dis)figuration: Sue Coe's *Dead Meat*

We find many faces in the paintings and drawings collected in Sue Coe's book *Dead Meat*, a collection of sketches, paintings, and drawings that Coe compiled over a six-year period while traveling to slaughterhouses and feedlots around North America (1995, v). Hundreds of faces, even thousands, perhaps. And we do not have to find them. They find us. They stare out at us on nearly every page, by turns fearful, afflicted, or innocent. What is remarkable here, though, is that the faces belong mainly to the animals—"livestock," so called. In fact, it is hard to find a human being with a face at all, and when we do find them, they are usually misshapen or contorted. How are we to understand this?

One way that suggests itself immediately, of course, is by means of the theorization of the ethics of "the face" in contemporary philosophy and theory—a debate that has conspicuously involved Emmanuel Levinas, Jacques Derrida, and Gilles Deleuze and Félix Guattari, among others. Levinas, as is well known, theorizes the ethical call of the face as the site of an unanswerable obligation to which I am held "hostage," to use his term, in an infinite responsibility to the other. As Derrida himself has observed, however, even though the subject is held hostage to the other by the first imperative of the intersubjective relation—"Thou shalt not kill"—in Levinas (as in the Judeo-Christian tradition generally), this is not understood as a "Thou shalt not put to death the living in general." For Levinas, the subject is "man," whose ethical standing is secured by his access to both *logos* and language and the Word, and so, as Derrida puts it, in Levinas the subject resides in "a world where sacrifice is possible and where it is not forbidden to make an attempt on life in general, but only on the life of man" (1991, 112–13).[2] For Derrida, on the other hand—in what has emerged over the past few years as a famous moment in his later writings—the animal "has its point of view regarding me. The point of view of the absolute other, and nothing will have ever done more to make me think through this absolute alterity of the neighbor than these moments when I see myself naked under the gaze of a cat" (2002, 380). And from the vantage of Deleuze and Guattari, Derrida's critique of Levinas here might be viewed as leaving intact a certain humanist schema of the scopic and the visual, which their critique of "faciality" in *A Thousand Plateaus* is calculated to dismantle in its insistence that "the face" is not a location, still less a body part, but rather a kind of "grid" or "diagram" that configures the space of intersubjective relations and desire itself, making them available only at the expense of "fixity" and "identity" (1987, 167–91). To put it schematically, Deleuze and Guattari might well ask of Derrida, "How can the moment of being looked at by his cat—not just 'naked' but '*seeing myself* naked under the gaze of a cat'—be divorced from the face?" How can the looking-back of the animal—and the ethical call harbored by that look—be disengaged from the humanism for which the face (and "faciality" generally) is perhaps the fundamental figure?[3]

Art historian and critic Michael Fried gives a rather different account from the Levinasian one of the question of the face in his 1987 book *Realism, Writing, Disfiguration: On Thomas Eakins and Stephen Crane*, where he offers an analysis of figuration and representation that I think will help shed light on the particularity of Coe's strategies and how we might assess their ethical force. The key point of contact with the motifs we have sounded out thus far, however briefly, is readily voiced in the title of the essay on Crane that makes up the second half of the book: "Stephen Crane's Upturned Faces," where the intense visuality of Crane's prose is also indexed to *the face*—and to the blank page as its double or stand-in—and its ethical call upon us. Pertinent here too is the fact that in Crane, in Eakins, and in Coe, we will be dealing with—immersed in, really—scenes of violence and responsibility, primarily war (as in Crane's *The Red Badge of Courage*), the surgical theatre of Eakins's great painting *The Gross Clinic*, and, of course, the killing floors of Coe's *Dead Meat*.

What Fried finds in Crane is "a mode of literary representation that involves a major emphasis on acts of *seeing*, both literal and metaphorical" (1987, 114). But what is usually called Crane's "impressionistic" style should instead be understood, Fried argues, as a remarkable plumbing of the relationship of "a primitive ontological difference between the allegedly upright or 'erect' space of reality and the horizontal 'space' of writing," which manifests itself in Crane as "an implicit contrast between the respective 'spaces' of reality and literary representation" (1987, 99). This difference is related to the extraordinary (and extraordinarily haunting and even uncanny) network of faces in Crane's fiction—primarily faces of the dead that stare back at us with unseeing eyes—by virtue of the requirement "that a human character, ordinarily upright and so to speak forward-looking, be rendered horizontal and upward-facing so as to match the horizontality and upward-facingness of the blank page" (ibid.). On the one hand (and here, I think, the connection to Coe's animal faces is quite clear), the faces of the dead—like the blank page—stare back at us and ask for our conferral of meaning, through representation, upon their abjection or suffering (this is rendered in an especially powerful way in Crane's fiction that deals with war, such as *The Red Badge of Courage*). But at the same time, as figures for the " 'unnatural' process" of writing itself—when "the upward-facingness of the corpse, hence of the page," is considered "not so much as a brute given [but] as a kind of artifact"—they are *products* of that very process of representation itself (1987, 100).

In trying to bring the reader/viewer face-to-face with the world through writing, however, the writer only succeeds in *de*facing the world or, to use Fried's term, *disfiguring* it. The dilemma in Crane is that the more he succeeds in this enterprise, the more he, in another sense, fails. This is so, Fried argues, because insofar as those "desemanticizing" aspects of Crane's writing (visuality, sonority, dialect, and manipulation of perceptual scale, just to name a few) do their job, they interpose themselves, in their own materiality, between the reader and the world that that "realist" project was supposedly intended to represent, so that the world (though he does not put it this way) almost becomes a "host," if you will, for an essentially "vampiristic" relationship to the writerly or representational project. As Fried puts it, "Wouldn't such a development threaten to abort the realization of the 'impressionist' project as classically conceived? In fact, would it not call into question the very basis of writing as communication—the tendency of the written word partly to 'efface' itself in favor of its meaning in the acts of writing and reading?" (1987, 119–20).

For Fried, this uncanny or vampiristic quality of Crane's style is symptomatic of Crane's need to performatively confront "the scene of writing" through "a mechanism of displacement," and "to do so in a manner that positively obscured the meaning of those representations from both writer and reader." "And this suggests," he continues,

> that the passages that describe the faces and recount responses to them are where Crane's unconscious fixation on the scene of writing not only comes closest to *surfacing* in a sustained and deliberate manner but also, precisely owing to the "manifestly" dreadful nature of the faces and of the vicissitudes that befall them, is most emphatically *repressed*. In other words, the thematization of writing as violent disfigurement and its association with effects of horror and repugnance but also of intense fascination allowed the writer, and *a fortiori* the reader, to remain unconscious of the very possibility of such a thematization. (1987, 120–21)

We are now in a position to begin to glimpse how different things are in Coe's handling of what we could call, after Fried, the scene of representation or figuration, whose index in both cases is a certain rendering—and in Crane's case, *rending*—of the face. We remember Fried's observation "that a human character, ordinarily upright and so to speak forward-looking, be rendered horizontal and upward-facing so as to match the horizontality and upward-facingness of the blank page" (1987, 99). In Coe, however, we find a double reversal of this dynamic. First, the violence that in Crane renders the human corpse horizontal and upward facing is in *Dead Meat* associated with a force that takes the "naturally" occurring horizontality of the animals portrayed (living, as they do, on all fours) and renders it strongly vertical—namely, in the endless rows and rows of hoisted, hanging animal corpses in the slaughterhouse and the packing plant. It is as if the animals cannot be allowed to assume the vertical, upright posture reserved (as even Freud tells us in *Civilization and Its Discontents*) for the human, without at the same stroke being *defaced*—in many cases, quite literally (i.e., beheaded).[4]

At the same time—a strict corollary by this logic—the slaughterhouse workers remain mired in a strongly horizontal plane and, not surprisingly, their faces are often "beastly" or "animalistic" in the traditional, speciesist sense of the word. The logic that systematically works its way through most of these pieces, then, is that the concrete, individual animal body—an individuality emphasized in pieces such as *Cow 13* and *Goat Outside Slaughterhouse* (Figure 6.1)—is, through a process of corporately organized Taylorization, mechanistically born, bred, killed, and dismembered in a process through which it comes to have meaning for the "carnophallogocentric" *socius* (to use Derrida's well-known term) only by being reconstituted as "meat" or "pork"—a semantic transformation and mystification that is itself paralleled by the material manifestation of identical, shrink-wrapped packages of brightly colored meat in the grocery store counter, now thoroughly dissociated from the reality of its material production (1991, 113).[5] And this systematic violence against the animals is itself doubled by a less brutal, though no less systematic, violence that

Figure 6.1. Sue Coe, *Goat Outside Slaughterhouse*

attends the workers who are forced by the nature of capitalism itself to do such work—a point graphically captured in Coe's rendering of the meat-packing workers in painting after painting.

Second, however—and this is the point I would like to emphasize—what we find here is not the "excruciated" relationship to representation that Fried emphasizes in Crane and Eakins but rather its apparent displacement onto forces external to the work of representation itself—forces whose effects the artwork registers and then intensifies. The violence we find here is not "artifactual" (associated with the inescapable violence and disfiguration of representation itself) but is instead associated with the external—that is, extrarepresentational—forces of capitalism and factory farming. We could say, in other words, that—in direct contrast to Fried's Crane—Coe's painting aspires to the condition of writing, but writing understood not as representation divided against itself—not as *différance* or *iterability*, to borrow Derrida's terms, which are invoked by Fried (1987, 163, n. 1, 185, n.28)—but rather as the direct communication of a semantic and, as it were, external content, of which the artwork is a faithful (or perhaps we should say "dramatic") enough representation to, indeed, didactically incite ethical action and change on the part of the viewer.

Yet precisely here an interesting problem manifests itself. While Coe is certainly within her rights to see the ethical function of (her) art, at least in one sense, as drawing our attention, as powerfully as possible, to the untold horrors of the slaughterhouse, on another level—and it is this level that will be handled with considerable sophistication, I think, in Eduardo Kac's work—that ethical function, and the representationalism upon which it depends, relies upon a certain disavowal of the violence—what Fried calls "disfiguration"—of representation itself, which immediately leads to a very obvious question we might ask of Coe: If the ethical function of art is what Coe thinks it is, then why not just show people photographs of stockyards, slaughterhouses, and the killing floor to achieve this end? To put it another way, what does art *add*? And what does it mean that her art has to be *more* than real to be real? Is not the "melodrama of visibility" (to use Fried's phrase) that we find in *Dead Meat*, which is calculated to "give the animal a face," also, in another sense, an *effacement* of the very reality it aims to represent, one that quite conspicuously manifests itself in the hyperbole, disfiguration, and melodrama of Coe's work? The paradoxical result for Coe's work, then, is that appeals to us to read it as directly (indeed, melodramatically) legible of the content it represents, but the only way it achieves that end, is *through* its figural excess, which is precisely *not* of the slaughterhouse but of the interposing materiality of representation itself.

We can unpack the implications of this point by remembering Fried's discussion of "what might be called a drama, some would even say a melodrama, of visibility" in Eakins's *The Gross Clinic* [Figure 6.2], which may be brought into sharp contrast with the very different "melodrama" we find in Coe's *Dead Meat* project (1987, 59).[6] My point here in calling Coe's work "melodramatic" is not that it exaggerates what really goes on in a slaughterhouse but, rather, that in Coe's work, *nothing is hidden* from us. On the contrary, the paintings seem to form a kind of theatre calculated to produce a "surefire effect" (to use Fried's characterization of "theatricality") by "playing to the audience," as the figures in the paintings—human and animal—repeatedly look out at us, imploringly, fearfully, or sadistically, as if the entire affair inside the space of the painting is staged *only* for us (1998, 40). Unlike the experience of the viewer in what Fried calls the "absorptive" tradition in painting that culminates in modernist abstraction, the viewer in Coe's work is not "denied," as Fried puts it, but rather addressed and held responsible, as it were, even culpable, for what is being shown inside of the frame.

Here—to return to *The Gross Clinic*—two conspicuous features of Eakins's painting noted by Fried are very much to the point: the rendering of the surgical patient's body, and the cringing figure of an

Figure 6.2. Thomas Eakins, *The Gross Clinic* (1875), oil on canvas, 96" x 78"

older woman, usually taken to be the patient's mother. As for the first, Fried notes that "the portions of the body that can be seen are not readily identifiable, so that our initial and persisting though not quite final impression is of a few scarcely differentiated body parts rather than of a coherent if momentarily indecipherable ensemble" (1987, 59). In fact, Fried likens this presentation to something like a dismembering, an act of "deliberate aggression" and even "sadism" that ultimately is an index of "the attitude toward the viewer that that rendering implies"— an especially intense version of the attitude typical of what Fried elsewhere famously calls the "absorptive" tradition in painting (ibid.). Similarly, the cringing figure dramatizes "the pain of seeing," in both "the emphatic *emptiness* of her clawlike left hand," the "violent contortion" of which is "apprehended by the viewer as a threat—at a minimum, an offense—to vision as such," and "the *sightlessness* that . . . she so feelingly embodies" (1987, 62). In these "aggressions," as Fried calls them, these gestures of "disfiguration," Fried finds in the painting "an implied affront to seeing," a "stunning or, worse, a wounding of seeing—that leads me to imagine that the definitive realist painting would be one that the viewer literally could not bear to look at" (1987, 64–65).

Here, I think, we get a very precise sense of the differences between the force of "disfiguration" at work in Eakins's representationalism and Coe's. For in Coe, although there is disfiguration aplenty, it is never a disfiguration that resists vision or interpretation—quite the contrary, it invites single, univocal reading. The violence of Eakins's "affront to seeing" that manifests itself in *The Gross Clinic* as incision, deformation, and even, in a sense, dismemberment (a violence displaced and contained by being thematized, as Fried notes, in terms of the "necessary" surgery being performed), is matched by the reverse dynamic in Coe. The almost nightmarish, infernal scenes of violence before us *hide nothing*, and for that very reason, the artist, as it were, has no blood on her hands. (*That* is reserved, of course, for the forces of capitalism and Taylorization referenced in the work's semantic content.)

In this light, we can sharpen our sense (if you will pardon the expression) of the difference between Coe's representationalism and Eakins's by reminding ourselves of the signifying force of the surgeon's scalpel in *The Gross Clinic* as glossed by Fried. If Eakins represents himself allegorically through the figure of Gross, then the scalpel serves to remind us—rather startlingly, even traumatically—that Eakins is "divided or excruciated between competing systems of representation"; on the one hand, the scalpel, "being hard and sharp, an instrument for cutting, belongs unmistakably to the system of writing/drawing," but on the other, because the scalpel is marked by an *outré*, almost three-dimensional drop of blood on its tip, it "refers, by means of an irresistible analogy," to the system of painting—almost as if the drop of blood *were* paint and the surgeon/painter carefully and dramatically deliberates its violent application (1987, 88). In this light, we might well say of Coe's *Dead Meat* that the knives and hooks of the slaughterhouse are *never* associated with the brush of the painter and the violence of representation-as-disfiguration. Thus if Eakins's putative "realism" in fact harbors a deeper, more unsettling antirealism or, perhaps better, *irrealism*, then Coe's hyperbolic, melodramatic renderings themselves harbor a more fundamental (and a more fundamentally comforting) representationalism, a signifying regime whose best name might well be "faciality"—*even if* that faciality is extended across species lines to include, even privilege (as if to somehow redeem their suffering), the nonhuman animals around which the paintings are built.

The opposite of this regime—or, more precisely, as Derrida would put it, that which remains "heterogeneous" to it, not its simple other—might well be figured in the network of a-signifying forms and their serial iteration that wends its way throughout the works collected in *Dead Meat* (1988, 116). Chief among these are the chains, hooks, tubes, belts, hoses, ducts, and the like that form a kind of ongoing cipher in the paintings, often extending beyond the borders of the pictorial space, suggesting their intrication in some larger insidious network (Figure 6.3)—a logic that also is extended to cover the representation of the masses of animal bodies themselves in pieces such as *Lo Cholesterol Buffalo* or *Feedlot*.

From Coe's representationalist point of view, of course, this network is directly associated with the force of capitalism, Taylorization, and the disassembly line they put in place. In the sense I am emphasizing here, however, we might see it as figuring instead a kind of displacement or domestication of what Derrida calls "iterability"—or, as Fried would have it, a kind of visible repression that traces and scores the otherwise representational logic of the paintings. This logic even extends, I would suggest, to the ubiquitous numbered ear tags that mark the animals as fodder in the larger machine of agribusiness and factory farming, with the sheer abstractness and pure seriality of the numerical system signifying nothing *except* this force. Here, the painting *Goat Outside Slaughterhouse* (Figure 6.1) is all the more striking in the contrast between the almost sculptural modeling of the animal's head and the abstract numbers of the contrasting ear tag, which are not only iterations of the same shape but which also in their form recall the network of figures I have just noted in pieces such as *Ham Scrubber* (Figure 6.3).

Given the conceptual coordinates of Coe's *Dead Meat* project, we can surmise that this force of seriality would eventually find its most extreme logical extension in genetic engineering and, beyond that, in cloning—an eventuality graphically depicted in Coe's painting *Future Genetics Inc.* Here again, however, we can interpret this in a second sense rather at odds with the artist's own. For while Coe's painting depicts the perverse extension of Taylorized factory farming to the production of misshapen and deformed animal mutants in a subterranean laboratory, there is another sense in which we may view this logic as endemic to *representation itself.* The clone may be "the image of the perfect servant, the obedient instrument of the master creator's will," as W. J. T. Mitchell puts it, but it also activates "the deepest phobias about mimesis, copying, and the horror of the uncanny double" (2005, 25). Or, to put this in Derrida's terms, the dream of pure, Taylorized seriality is repetition *without* difference, but the very meaning and force of iterability is that repetition—and representation—can only take place in and through the potentially mutating work of difference, the specific material, embodied, pragmatic instance that threatens any dream of purity, always shadowing pure seriality with the uncanny referenced by Mitchell. And this opens up a second ethical register around the question of representation and its logic—one quite different from what Coe has in mind—that harbors very real stakes for how we understand the human/animal relation.

Figure 6.3. Sue Coe, *Ham Scrubber*

For as Derrida has argued, the constitutive fantasy of humanism is that the human separates itself from the rest of the domain of "the living" by alone escaping subjection to the deconstructive force of iterability and the trace that in fact extends to *all* forms of representation and signification, not just its paradigmatic case, language—a point amplified powerfully in an important late essay on Lacan and the animal.[7] In this second ethical register, the critique of speciesism *emerges*, in fact, from the critique of representationalism along the lines traced by Derrida in "Eating Well," where he suggests that

> if one reinscribes language in a network of possibilities that do not merely encompass it but mark it irreducibly from the inside, everything changes. I am thinking in particular of the mark in general, of the trace, of iterability, of *différance*. These possibilities or necessities, without which there would be no language, are themselves not only human. . . . And what I am proposing here should allow us to take into account scientific knowledge about the complexity of "animal languages," genetic coding, all forms of marking within which so-called human language, as original as it might be, does not allow us to "cut" once and for all where we would in general like to cut. (1991, 116–17)

This may seem a very different kind of "cut" from the ones we witness in *Dead Meat,* but in fact, Derrida suggests, the "sacrificial symbolic economy" of "carnophallogocentrism" that subordinates woman to man and nonhuman animals to both is directly related to—even moti-vates—what we witness in Coe's work. "The subject does not want to just master and possess nature actively," Derrida writes. "In our cultures, he accepts sacrifice and eats flesh. . . . [I]n our countries, who would stand any chance of becoming a *chef d'Etat* (a head of state), of thereby acceding "to the head," by publicly, and therefore exemplarily, declaring him- or herself to be a vegetarian?" (1991, 114).

The More You Look, The Less You See: Eduardo Kac

In October 2001, Eduardo Kac presented his project *The Eighth Day* (Figure 6.4) in a gallery at Arizona State University, on the heels of what is probably his most famous undertaking, *GFP Bunny* (2000) (Figure 6.5). Here again, Kac uses "transgenic" life-forms (in this case, mice, zebrafish, tobacco plants, and a colony of amoebae, instead of a rabbit) modified by introducing into them an enhanced GFP (green florescent protein, derived from the jellyfish *Aequorea Victoria*) gene that makes them glow

Figure 6.4. Eduardo Kac, *The Eighth Day*

Figure 6.5. Eduardo Kac, *GFP Bunny*, 2000. Alba, the fluorescent rabbit

green under certain lighting conditions. As in that earlier work, however, GFP life-forms are only part of the story (2005, 66). In *The Eighth Day,* viewers enter a dark space with a glowing, blue-lit Plexiglas semisphere at its center, surrounded by the sounds of waves washing ashore. Inside the terrarium are the life-forms just mentioned, as well as a specially designed "biobot," which contains as its "cerebellum" the GFP amoebae. When the amoebae move toward one of the six legs of the biobot, their movement is tracked by a computer, which makes that particular leg contract. The biobot also serves as an "avatar," as Kac puts it, of Web participants, who can remotely control its "eye" with a pan and tilt actuator, so that "the overall perceivable behavior of the biobot is a combination of activity that takes place in the microscopic network of the amoebae and in the macroscopic human network." Meanwhile, viewers in the gallery can see the terrarium from both inside and outside the dome, by means of access to a Web interface installed in the gallery space, which includes, in addition to a biobot view, a feed from a "bird's-eye view" camera installed above the dome (2005, 291–92).

When we leave behind the technical and logistical aspects of the piece (which are considerable) to address the work's intellectual, ethical, and social implications, we enter another order of complexity. Arlindo Machado's comments in the collection of essays that accompanied *The Eighth Day* are fairly representative of these discussions. He writes:

> Transgenic forms of life are often stigmatized for being produced in the laboratory, in part because of the economic (and possibly warlike) interests that motivate their creation. It is almost inevitable that nontechnical discussions involving biotechnologies take on a conservative bias, recalling scenarios of apocalyptic science fiction or even dogmatic interdictions of religious order. . . . The more experimental and much less conformist sphere of art—with its emphasis on creation, by means of genetic engineering, of works which are simply beautiful, not utilitarian or potentially profit making; along with the relocation of genetically modified products in "cultural" spaces such as museums and art galleries, or in public spaces, or even in homes . . . —all this could help to elevate public discussion of genetics and transgenics to a more sophisticated level. (2003, 94–95)

This is essentially the thrust as well of Kac's own "manifesto" on transgenic art, but the artist takes the additional step there of insisting that "artists can contribute to increase global diversity by inventing new life-forms," and he imagines a day in the not-too-distant future when "the artist literally becomes a genetic programmer who can create life-forms by writing or altering a given (genetic) sequence" (2005, 237, 243).

This insistence, of course, complicates an already complicated situation considerably, because it invites the sorts of trepidations rightly raised by critics such as Steve Baker, who writes that Kac "engages with the animal through techniques that strike many people as meddlesome, invasive, and profoundly unethical" (2003, 29). It is not that any of the animals used in his work are harmed (they are not, and Kac has repeatedly made it clear how seriously he takes his responsibility for the care and well-being of the animals involved), but rather that "Kac seems to overlook the larger picture," as Baker puts it, namely, that his work depends upon and in a fundamental sense reproduces an entire set of institutions and practices of scientific research subjecting millions of animals a year to distressing, often painful, and usually fatal experimentation, a subjection of nonhuman beings of "unprecedented proportions," as Derrida puts it, in which "traditional forms of treatment of the animal have been turned upside down" and replaced by "an artificial, infernal, virtually interminable survival, in conditions that previous generations would have judged monstrous." In fact, Derrida flatly (and accurately) calls it a "hell" of "experimentation," a massive and systematized "violence that some would compare to the worst cases of genocide" (qtd. in Baker 2003, 34–35).

Such concerns are very important, of course, but I do not want to pursue them any further here—in part, because they have received ample airtime in the discussions of Kac's work, but primarily because certain habitual oversimplifications endemic to addressing those concerns have tended to mask crucial aspects of Kac's work, features that have a less obvious and, as it were, thematic relation to how his projects ethically intervene in our received views of the human/animal relationship and, beyond that, in the question of posthumanism generally.

Something of the different direction I want to pursue is evoked by Kac early on in the "Transgenic Art" manifesto, where he writes, "More than making visible the invisible, art needs to raise our awareness of what firmly remains beyond our visual reach but, nonetheless, affects us directly. Two of the most prominent technologies operating beyond vision are digital implants and genetic engineering" (2005, 236). In a recent essay on art and human genomics, critic Marek Wieczorek extends the point when he asks, "How do we picture a new age of genetic manipulation . . . a literal synergy between computing and biology?" This is not just a question of *representation* in any straightforward sense, because "the digital code of the genome, emblematic of a new mode of consciousness," is "not a spatial blueprint of life, not a two-dimensional plan of what a heart or liver looks like, but a long string of nucleotides written in endless permutations" (2000, 59). And what this means, in turn, is that the problem of picturing this immense revolution "may not simply be a matter of new *forms* of visuality," but rather "reconciling form with principle" (60).[8]

Here—and this is a rather different understanding altogether from what we find in Fried—Wieczorek finds a precursor to this new work of Kac's regarding the parallels between art and scientific theory in minimalism, with its "potentially endless sequence of repeated shapes." Just as "digitally encoded information has no intrinsic relationship to the form in which it is decoded"—"it is not tied to a singular, inherently meaningful form"—so in minimalism "repetition replaces singularity." Moreover, in minimalism, "art acknowledges the viewer, whose physical interaction with the work produces ever-shifting viewpoints over time, through a kind of feedback loop," which parallels a similar emphasis in cybernetics on what Humberto Maturana and Francisco Varela call the "autopoiesis" (or Niklas Luhmann will call the "self-reference") of the observing system—a fact Kac's work insists on again and again, most obviously in his inclusion in the work itself of remote, Internet-based observer/participants. Here, however, the point is not, as Wieczorek puts it, that "reflexivity is regressive," much like the "obsessively pointless variations of LeWitts' incomplete open cubes or Judd's boxes" (2000, 59–60). Rather, the point, I think, is that it is *recursive*—it uses its own outputs as inputs, as Luhmann defines it (1990, 72). And it is only on the basis of that recursivity—a dynamic process that takes *time*—that reflexivity becomes *productive* and not an endlessly repeating, proverbial "hall of mirrors" associated with the most clichéd aspects of postmodernity.

There are two points here, one logical and one, as it were, biological. As for the first, Wieczorek captures something of how Kac's work thematizes the central fact—a logical and cognitive fact—about recursive self-reference as Luhmann has theorized it, namely, that observation (precisely because it is contingent and self-referential) will always "maintain the world as severed by distinctions, frames, and forms," and this "partiality precludes any possibility of representation of mimesis and any 'holistic' theory." Thus, Luhmann writes, "The world is observable *because* it is unobservable" (1995, 44, 46).

Of more immediate relevance for Kac's work, however, is the second point, the biological one: that recursive self-reference is crucial to how different kinds of autopoietic beings establish their *difference* from everything else in the world, which is to say their specific ways of *being* in the world—a "being" that is now thoroughly subordinated to an autopoietic *becoming*. For Kac—and here is where Wieczorek is right that it is not simply a matter of new *forms* of visuality—this calls for a recalibration, redistribution, and displacement of the relationship between meaning and the entire sensorium of living beings, in which visuality itself—as the human sensory apparatus par excellence—is now thoroughly decentered and subjected to a rather different kind of logic.

To put it another way, Kac subverts the centrality of the human and anthrocentric modes of knowing and experiencing the world by displacing the centrality of its metonymic stand-in, human (and humanist) visuality. He does this in several different ways, some of which are comparatively straightforward, such as *Darker Than Night* (1999) and *Rara Avis* (1996). In the former, the viewer is linked in a communicational loop to roughly 300 fruit bats via a "batbot" implanted in their cave, which enables the viewer to "hear" the converted echolocation sonar signals of the living bats, while the viewer wears a VR headset that converts the batbot's sonar emissions into a abstract visual display (2005, 202–203). In the latter, viewers donned a headset linked to a camera in the head of a large robotic bird in an enclosure, surrounded by living birds, which enabled the viewer to look out from the robotic bird's point of view. In both works, sounds (*Rara Avis*) and sonar signals (*Darker Than Night*) originating from human participants are reintroduced into the animals' environment, allowing them to experience the presence of an absent, human other (162–66).

More interesting still, I think, is how Kac's work also exploits what we might call our lust for the visual and its (humanist) centrality by trading upon it repeatedly (the glow-in-the-dark creatures, the rather *outré* coloring of the bird in *Rara Avis*, or even the playful visual pun on the human eyeball in *Teleporting an Unknown State* [1996] [Figure 6.6], just to name a few). This is not just, I think, a matter of the "scopic reversal" that is a "recurring theme" in Kac's work (particularly the works on telepresence), nor is it just about a "dialogical interchange" that serves "to multiply the 'points of view' available," as in *The Eighth Day*. Nor is it exactly that "to the extent that something living—particularly a mammal—glows green, we have an index of alterity" (an interpretation resisted by Kac, by the way) (Collins 2003, 99).[9]

In fact, I would argue that the use of GFP in Kac's work, particularly with the rabbit Alba in *GFP Bunny*, operates as a kind of feint or lure that trades upon the very humanist centrality of vision that Kac's work ends up subverting. On display here, in other words, are the humanist ways in which we produce and mark the other (including the animal other), our "carnophallogocentric" visual appetite, displayed here in the form of spectacle, which gets "fed" in this instance by GFP. From this vantage, the point is perhaps not so much, as W. J. T. Mitchell puts it

Figure 6.6. Eduardo Kac, *Teleporting an Unknown State*, 1994–1996

in his widely read essay "The Work of Art in the Age of Biocybernetic Reproduction," that "Kac's work dramatizes the difficulty biocybernetic art has in making its object or model visible," because "the object of mimesis here is really the invisibility of the genetic revolution, its inaccessibility to representation" (2005, 328). Rather, it is that Kac's work—with its glow-in-the-dark creatures and its black lights, drawn as much from the storehouse of cheesy mass culture as anywhere—makes all of this *all too* visible by eliciting and manipulating very familiar forms and conventions of contemporary visual appetite. And in doing so, it may be understood against the backdrop of Mitchell's larger point about the work of art in an age of biocybernetic reproduction: that the "curious twist" of our moment is that "the digital is declared to be triumphant at the very same moment that a frenzy of the image and spectacle is announced" (315).

It is a question, then, of what we might call the "place" of the visual—but, eventually, for that very reason, of everything else too. And this involves in Kac's work—as one might now expect—a circular and indeed a recursive procedure, where the artist uses or otherwise appeals to specifically human visual habits, conventions, and so on for the purposes of making the point that the visual, as we traditionally think of it, can precisely no longer be indexed to those conventions and habits at all. In this light, one way to underscore the difference between productive recursivity in Kac's work and a mere "hall of mirrors" reflexivity is to say that the whole point of the glow-in-the-dark rabbit of *GFP Bunny* and how it seizes upon certain spectularizing modes of human visuality is that *the harder you look, the less you see.* Alba's "meaning," if we want to put it that way, is not to be found in the brute fact of the glow of her coat; in fact, one might well say the "meaning" of the work is everywhere *but* there.[10]

From this vantage, we might well think of the strategy Kac deploys in the work *Time Capsule* (1997), as framed by this same logic. In that piece, Kac was televised and simultaneously Web cast injecting into his leg a tiny microchip with a unique identification number that reveals itself when scanned—a device commonly used for registering and recovering companion animals. As part of the work, Kac registered himself in an "Identichip" database as both "animal" and "owner." In addition, the work included seven sepia-toned photographs of members of Kac's family from previous generations and a telerobotic Web scanning and displayed X-ray of the implant in Kac's leg. Here again, Kac's deployment of spectacle and the visual generally makes the point, I think, that the significance of the work is everywhere *except* in its elements of visuality and spectacle. It begins to dawn on us just how true this is when we understand, as Edward Lucie-Smith points out, that Kac is of Jewish origin, that a number of his family members (some of them pictured in the sepia-toned photographs) were Polish Jews who died in the Nazi Holocaust, and how "the microchip incorporating a number alludes to the numbers tattooed on the arms of those who were herded into concentration camps"—but here, of course, the identifying numbers cannot be read (2003, 22). "Herded" is indeed a word to be insisted upon here, as this piece also focuses our attention not on livestock animals but *domestic* animals—mainly cats and dogs—for whom the chip is designed, animals that a vast majority of owners describe as family members. Are they less "animals" than those other living beings we call "meat"? Than the Jews in the eyes of the Nazis who forced them into cattle cars at gunpoint? Moreover, this welter of complicated associations and category crossings can be amplified one last time when we remind ourselves of Derrida's characterization of contemporary forms of animal exploitation in biomedical research and factory farming as a "holocaust."

All of this changes completely, I think, the understanding of "theatricality" as criticized by Fried. The point is not just, as Fried would have it, that Kac's work is "theatre" (which, in his terms, it surely would be), but rather that "theatre" is not doing the work Fried thinks it does. In Kac, the artwork does indeed "play up" to the viewer, but only, as Derrida would put it, to lead the viewer to the realization that the only place the meaning of the work may be found is no place, not where the viewer irresistibly looks (e.g., at the spectacle of the glow-in-the-dark creatures) but rather precisely where the viewer *does not* see—not *refuses* to look, or, even, is *prevented* from seeing but rather *cannot* see. If we keep in mind that theatricality depends first and foremost upon spatial distribution, then we can appreciate the resonance of the following comment by Derrida for Kac's attempt—and the ethics of that attempt—to *situate* the visual in ways that fundamentally trouble how we have typically indexed the (human) animal sensorium to the human/animal ontological divide. "I am not sure that space is essentially mastered by [*livré à*] the look," Derrida remarks; "space isn't only the visible, and moreover the invisible"—an invisible that is itself "not simply the opposite of vision" (1994, 24).[11] In this light, we can see more clearly—or perhaps, I should say, more "obliquely"—how Kac's theatricalization of visuality does not evade the viewer's "finitude" and "humanness" (as Fried would have it) but rather *underscores* it in the specifically posthumanist sense that the field of meaning and experience is no longer thought to be exhausted by the self-reference of a particularly, even acutely, human visuality (1998, 42).

In the end, then, the contrast between Coe and Kac helps us see, in the realm of art, the difference between two different kinds of posthumanism: a humanist posthumanism and a posthumanist one. Coe may be viewed as a posthumanist in the obvious and, as it were, thematic sense that she take seriously the ethical and even political challenges of the existence of nonhuman animals (this latter, in her cross-mapping of the exploitation of animals and of workers in factory farming within a more or less Marxist frame). But as I have argued elsewhere in an analysis of animal rights philosophy, one can well be committed to this posthumanist question in a humanist way—that is, in a way that reinstalls a very familiar figure of the human at the very center of the universe of experience (in animal rights philosophy) or representation (in Coe's work)—a subject who *then*, on the basis of that sovereignty, extends ethical consideration outward toward the nonhuman other.[12] In this light, Coe's work is humanist in a crucial sense, indeed, in the only sense that turns out to be absolutely fundamental to her work *as art*: that it relies upon a subject from whom *nothing, in principle, is hidden*. A subject, who if blind, is blind not constitutively (as I think Kac's work dramatizes in multiple ways), but only because he (and I would insist on the pronoun in this instance, for reasons Derrida makes clear)—has *not yet seen* what Coe's art is calculated to reveal so powerfully and clearly, so well *can* that subject see. And this complicates considerably—one might even say fatally—Coe's conception of art as a form of "witnessing" (1995, 72), for what must be witnessed is not just what we can see but also what we cannot see—*that* we cannot see, that only the other can see. That too must be witnessed. But by whom?

Notes

1. For the most comprehensive overview we have of the animal in contemporary art, see Baker 2000.

2. I have discussed the Derrida/Levinas difference on this point in some detail in my essay (2003b) "In the Shadow of Wittgenstein's Lion: Language, Ethics, and the Question of the Animal," esp. 17–18, 23–28. Also available as the second chapter of Wolfe (2003a) *Animal Rites,* 44–94.

3. See also Wolfe 2003a, 227–28, n. 1, and Lingis 2003.

4. See my discussion of this point in Freud in Wolfe 2003a, 2–3, 108–16.

5. For a by-now classic analysis of this process of production and renaming, which is also concerned with its phallocentrism (though not in Derrida's sense of the term), see Adams 1990.

6. I use the term *melodrama* in the specific sense that has some centrality to Fried's body of criticism and its fundamental contrast of "theatricality" (or "literalism") and "opticality" (or "absorption").

7. I refer to "And Say the Animal Responded?," esp. 137–38 in Wolfe 2003b, 121–46. See in particular 137–38.

8. A misrecognition of which by artist Thomas Grunfeld has led several critics—rightly, to my mind—to find his well-known taxidermy "hybrids" of body parts from different animals to be a radically unsatisfactory way of addressing this question. See also Lucie-Smith's (2003, 23) criticism in "Eduardo Kac and Transgenic Art."

9. For Kac's resistance, see the section "Alternatives to Alterity" in his essay "GFP Bunny," in *Telepresence and Bio Art,* 273–75.

10. This is directly related not only to the general point that Kac's work is to be viewed against the background that immediately precedes it (namely, conceptual art) as Mitchell (2005, 328) notes, but also—in a rather different sense—to Luhmann's insistence that the meaning of any work of art cannot be referenced, much less reduced, to its phenomenological or perceptual basis. On this point, see *Art as a Social System,* ch. 3, "Medium and Form."

11. See also Derrida's fascinating discussion of hearing versus seeing in the context of Aristotle's *De Anima* in *Eyes of the University,* 130–32.

12. See Wolfe 2003a. ch. 1.

Works Cited

Adams, Carol J. 1990. *The Sexual Politics of Meat.* New York: Continuum.
Baker, Steve. 2000. *The Postmodern Animal.* London: Reaktion.

———. 2003. "Philosophy in the Wild?" In *The Eighth Day: The Transgenic Art of Eduardo Kac*, ed. Sheilah Britton and Dan Collins, 27–38. Tempe: Institute for Studies in the Arts, Arizona State University.

Coe, Sue. 1995. *Dead Meat*. Introduction by Alexander Cockburn. Foreword by Tom Regan. New York: Four Walls Eight Windows.

Collins, Dan. 2003. "Tracking Chimeras." In *The Eighth Day*, 96–102.

Deleuze, Gilles, and Félix Guattari. 1987. *A Thousand Plateaus: Capitalism and Schizophrenia*. Trans. Brian Massumi. Minneapolis: University of Minnesota Press.

Derrida, Jacques. 1988. *Limited Inc*. Trans. Samuel Weber et al. Ed. Gerald Graff. Chicago, IL: Northwestern University Press.

———. 1991. " 'Eating Well,' or, the Calculation of the Subject." In *Who Comes After the Subject?*, ed. Eduardo Cadava, Peter Connor, and Jean-Luc Nancy, 96–119. New York: Routledge.

———. 1994. "The Spatial Arts: An Interview with Jacques Derrida," conducted by Peter Brunette and David Wills, trans. Lauri Volpe. In *Deconstruction and the Visual Arts*, ed. Peter Brunette and David Wills, 9–32. Cambridge: Cambridge University Press.

———. 2002. "The Animal That Therefore I Am (More To Follow)." Trans. David Wills. *Critical Inquiry* 28:2: 369–418.

———. 2003. "And Say the Animal Responded?" In *Zoontologies: The Question of the Animal*, ed. Cary Wolfe, 121–46. Trans. David Wills. Minneapolis: University of Minnesota Press.

———. 2004. *Eyes of the University: Right to Philosophy, Vol. 2*. Trans. Jan Plug et al. Stanford, CA: Stanford University Press.

Fried, Michael. 1987. *Realism, Writing, Disfiguration: On Thomas Eakins and Stephen Crane*. Chicago, IL: University of Chicago Press.

———. 1988. *Art and Objecthood: Essays and Reviews*. Chicago, IL: University of Chicago Press.

Kac, Eduardo. 2005. *Telepresence and Bio Art: Networking Humans, Rabbits & Robots*. Foreword by James Elkins. Ann Arbor: University of Michigan Press.

Lingis, Alphonso. 2003. "Animal Body, Inhuman Face." In *Zoontologies: The Question of the Animal*, ed. Cary Wolfe, 165–82. Minneapolis: University of Minnesota Press.

Lucie-Smith, Edward. 2003. "Eduardo Kac and Transgenic Art." In *The Eighth Day*, 20–26.

Luhmann, Niklas. 1990. "The Cognitive Program of Constructivism and a Reality that Remains Unknown." In *Self-organization: Portrait of a Scientific Revolution*, ed. Wolfgang Krohn et al., 64–85. Dordrecht: Kluwer.

———. 1995. "The Paradoxy of Observing Systems." *Cultural Critique* 31 (Fall): 37–55.

———. 2000. *Art as a Social System*. Trans. Eva M. Knodt. Stanford, CA: Stanford University Press.

Machado, Arlindo. 2003. "Towards a Transgenic Art." In *The Eighth Day*, ed. Sheilah Britton and Dan Collins. Tempe, AZ: Institute for Studies in the Arts, Arizona State University: 87–95.

Mitchell, W. J. T. 2005. *What Do Pictures Want?: The Lives and Loves of Images.* Chicago, IL: University of Chicago Press.

Wieczorek, Marek. 2000. "Playing with Life: Art and Human Genomics." *Art Journal* 59:3 (Fall): 59–60.

Wolfe, Cary. 2003a. *Animal Rites: American Culture, the Discourse of Species, and Posthumanist Theory.* Foreword by W. J. T. Mitchell. Chicago, IL: University of Chicago Press.

———. 2003b. "In the Shadow of Wittgenstein's Lion: Language, Ethics, and the Question of the Animal." In *Zoontologies: The Question of the Animal*, ed. Cary Wolfe, 1–57. Minneapolis: University of Minnesota Press.

Zizek, Slavoj. 1994. *The Metastases of Enjoyment: Six Essays on Woman and Causality.* London: Verso.

7

"They're There, and That's How We're Seeing It"

Olly and Suzi in the Antarctic

Steve Baker

Early in 2005, British artists Olly and Suzi traveled to Antarctica for the first time to spend three weeks making a visual record of their impressions of wildlife in the region, drawing everything from krill to leopard seals and jellyfish to penguins (Figure 7.1). Seeing themselves, broadly speaking, as environmentally conscious artists, Olly and Suzi (who, like Gilbert and George, are known professionally only by their first names) have worked collaboratively since the late 1980s. They are best known for painting and drawing endangered predators in their natural habitat at the closest possible quarters—whether tarantulas and green anacondas in Venezuela, wild dogs in Tanzania, or great white sharks underwater off the coast of Capetown. The two of them work simultaneously on each piece, "hand over hand" as they put it, in conditions that can be both inhospitable and dangerous, with the aim of conveying as directly as possible not only the beauty of these creatures but also the extent to which their lives and habitats are threatened.

In 1993 they abandoned studio work altogether, and the following ten years of worldwide expeditions to make artwork in the wild are extensively documented in their 2003 artists' monograph *Olly & Suzi: Arctic, Desert, Ocean, Jungle*. The fact that each of them had a young family by the time of its publication led them to decide to limit the number of expeditions undertaken each year and to explore the scope for making a certain amount of studio work again. This had to be done, however, in a manner that would not compromise their commitment to bringing to life the directness of their own experience of remote habitats in the eyes and minds of their viewers. This chapter attempts to chart

Figure 7.1. Olly and Suzi, *Penguins Sliding*. Caran Dache pencil on paper, Antarctica, 2005. Courtesy of the artists.

and contextualize this recent shift in their thinking, paying particular attention to the artists' own account of how they have put this new approach into practice.

Looking at Animals

There has been a recent renewal of interest in the topic of humans looking at nonhuman animals, with writers from a number of disciplines seeking to think more clearly about what is involved in that looking. Garry Marvin has made the important point that John Berger's classic (and still widely cited) essay "Why Look at Animals?" not only "approaches the visual encounter between humans and animals as though this were a mono-visual or one-dimensional process between the subject and the object," but also that it neglects to consider *how* people look at animals by focusing only on the question of "why" they do so (2005, 3).

Marvin's point is that there are many ways of looking, and he proposes "a continuum from unengaged to engaged" looking, briefly working his way through the idea of seeing, looking, watching, and observing nonhuman animals. He is quite explicit that his concern, and his emphasis, is on "the direct experience of the empirical animal rather than encountering an image" of it, though he also argues that from his anthropological perspective "a squirrel is represented the moment when recognized by us as a squirrel" (2005, 5, 6).

In terms of thinking about representations of animals—the concern of this chapter, and not of Marvin's—there are two difficulties with his argument. One is the implication that to represent is necessarily to classify, to judge, and to narrow rather than to open up human understanding of or engagement with the animal. It is a view with which most contemporary artists, and certainly Olly and Suzi, would find themselves out of sympathy. The other difficulty is that the terminology of looking itself differs according to whether the looked-at thing is the animal or an image of that animal. *Watching* a squirrel may indeed involve "a more attentive viewing" than merely looking at it, and observing may be a more "concentrated, attentive viewing guided by a particular interest" in the animal (5), but watching and observing are not terms that can usefully be applied to the viewing of drawings or paintings of animals—as if these flighty images might somehow vanish from the gallery wall if the viewer's attention was momentarily distracted. Notions of attention and attentiveness to animals and their habitats will be central to this account of Olly and Suzi's operation as artists, but they will call for a rather different vocabulary.

In the Introduction to this book, Sidney I. Dobrin and Sean Morey are specifically concerned with the use of images (including images of animals) "to create alternative ways of seeing nature and environment"— ways of seeing that might, broadly speaking, have a greater ecological integrity than much contemporary mass media imagery. However, in proposing the study of a visual eco-language that may have discernible "semiotic rules" Morey (in chapter 1, "A Rhetorical Look at Ecosee") perhaps underestimates the importance for contemporary artists of working in a more exploratory manner that is neither rule-bound nor particularly language-like.

Even Cary Wolfe's chapter in this book (chapter 6), which does directly explore the work of two contemporary artists, seeks to pin down the forms of their engagement with visual experience in terms that suggest that the visual representation of animals might sometimes be considered independently of questions of form, medium, and intention. He asks, for example, that given what Sue Coe sees as the ethical function of her own art (a furthering of the cause of animal rights), "Why not just show people photographs of stockyards, slaughterhouses, and the killing floor to achieve this end? To put it another way, what does art *add*?" These are very important questions, and ones that in a sense set the agenda for the detailed discussion of Olly and Suzi's work that follows here. The contention of this chapter, however, will be that productive answers may only be found in the particularities of an artist's practice, and that attentiveness to the form of that practice may make it difficult or even counterproductive to try to draw more general conclusions from it.

Picturing Animals

It is perhaps worth explaining at this stage that Olly and Suzi's own views about their work often are expressed with a disarming directness and simplicity: they do not see it as their role to theorize their practice. In what follows here, no particular effort will be made to establish a critical distance from those views, because it will be argued that the distinctiveness of their practice can in certain respects best be understood by taking seriously the manner and the terms in which they choose to describe it.

In the author's interview with the artists in August 2005, Olly reiterated a point on which they have insisted throughout their career: "Working in the wild is the very core of what we do." Speaking of the work made in their studio since 2003, including the *New Elements* series, shown at the Briggs Robinson Gallery in New York in 2005, he insisted that the work "is nothing without the experience, it is nothing. We cannot make this stuff unless we've been there."[1] Asked about how they would describe themselves, Suzi answered: "Maybe we're trying to be messengers." This emphasis on *being there*—which undoubtedly includes a continuing echo of Romanticism's concern with authentic experience—is therefore very much tied up with the idea of bringing something back in the form of a message of importance: a message about endangered wildlife, endangered habitats, the connectedness of things, and the beauty that will be lost if environmental degradation is allowed to continue.

The particular visual form of that message, however, is everything. Suzi articulated this rather clearly, contrasting their experience of working in the wild *as artists* with her own experience of a recent family holiday in Africa that had included some typical wildlife tourism. Keen to show her own family some of the big game in the area, they had set off in a jeep in the early hours of the morning, ticking off sightings of various species, until "we drove up next to this rhino, and it looked at us, and we looked at it, and I took a few photos, but I was in such a different frame of mind," she recognized at the time. "I suddenly realized quite how difficult what we do is. The *urgency* that I have with Olly when we're in the bush," she reflected, includes everything down to the mundane panic of "who's-got-the-sketchbook-and-where-are-the-pencils" in that "intense moment" when the animal fleetingly appears. "It's where we live," remarked Olly. Suzi added: "Yes, and it's just really brilliant."

The materials matter, not because pencil and paper are superior or more authentic than the photographic record, but because they do a different kind of work. Throughout their career, these artists have used both. On most of the trips documented in the artists' 2003 monograph, photographer Greg Williams—who has a successful photographic career in his own right—had traveled with them to document their interactions with animals in their natural environments, principally in photographs but also in film footage. Some of their most striking and best-known images—not least that of the great white shark surfacing to take a bite out of the artists' painting of that very creature (an image reproduced as a huge banner outside of London's Natural History Museum at the time of their 2001 exhibition there, as well as on the cover of the present author's 2000 book *The Postmodern Animal*)—are in fact photographs taken by Williams. Both Suzi and Olly are now individually taking some photographs themselves as part of their collaborative work, further complicating any answer to Cary Wolfe's question: What does art *add* to what the photograph might adequately convey on its own?

In that sense, at least, the question is something of a distraction. Like many contemporary artists with environmental concerns, who necessarily make use of local guides, assistants, and expertise on their various expeditions, and whose subject matter is recorded in a variety of media, Olly and Suzi make no hard-and-fast distinction between images that count as their "art" and other images that are merely supporting documentation. Asked about how decisions were made about which pieces fell within the scope of their collaborative work, Suzi explained that as far as Greg Williams's photographs were concerned, the three of them "all edited them together." Similarly, in relation to their own recent photographs, she and Olly "edit them together" in order to decide "whether it's his one or my one that says what we want it to say, so it's actually pretty similar to the drawing and painting, in that if he's done a better line, or the line that describes the back of a bear better than I have, then we go with that line." The remark gives a sense of the fluidity with which the artists move across and between these media.

Editing is becoming increasingly important for them, not least in terms of how the work might most effectively be displayed. Of the work from the Antarctica trip, Olly commented: "I really like this body of work; we've edited it down even more, and really as a body of work on Antarctica they need to be seen and shown together. It works best in context, with the photography, with the film, and then it'll all make sense." Suzi added that with any single drawing or painting "it's never really going to show the enormity of the situation," and that despite the practical and financial difficulties of holding work back from sale, the real advantage of displaying the work from a particular trip together is the chance "to give people the biggest picture that we can of what we saw." In fact, as will become clear, what Olly and Suzi hope to convey through their work is more than simply what they saw, as though they could somehow confine their experience to that of impartial, uninvolved observers.

"Looking So Hard You Can Hear Your Heart"

In Antarctica, although theoretically protected against commercial exploitation, it is the environment itself that is endangered, not least because of a significant increase in the fishing of the krill constituting a vital part of the food chain for other animals, right up to the ten-foot-long leopard seals that the artists were most dramatically to encounter on that trip (Figure 7.2). The leopard seals are predators that Olly described as being as imposing and intimidating as the polar bears that he and Suzi had encountered on an earlier Arctic expedition.

Figure 7.2. Olly and Suzi with George Duffield, *Leopard Seal*. Custom framed C-type photograph, Antarctica, 2005. Courtesy of the artists.

On the Antarctic trip on the Pelagic Australis (a seventy-five-foot aluminum hull Antarctic sailing yacht with a crew of three), Olly and Suzi were working alongside underwater photographer George Duffield—taking the place of Greg Williams on this occasion—and the wildlife filmmaker Doug Allan, who was operating both as expert guide and underwater cameraman filming material for a documentary on Olly and Suzi's work.[2] Anchored off of Hovgard Island, the dives they made in order to find, paint, film, and photograph the region's solitary leopard seals were the most trying part of the expedition. Not least of the problems was the intensity of the cold, and the artists' ill-advised decision to try out so-called "dexterous" gloves to aid the process of painting underwater—gloves that proved wholly inadequate for the job.

Talking to them subsequently about *what it actually felt like* to be working underwater in that environment drew out some of their most revealing responses. Despite the intensity of their physical discomfort, and concern for their fingers ("we were in a lot of trouble," Olly acknowledged), they seem not to have lost their sense of purpose. As Suzi explained:

> At that moment it feels like I've come all this way, all these things have happened for us to be here, and I may be really cold, or my ears may be hurting. . . . I remember one dive we did and my ear was killing me. Olly was going, you know, pointing up, do I want to go up? And my reaction was no, I want to stay down, I want to get this done, because we're here, and it's all happening *now*, so we've got to do it, otherwise what have we come all the way out here for? And I think that feeling often overrides everything else at that moment.

The whole situation was complicated by their having to take into account Doug Allan's need to document their work and their interactions with the seals for the film. In particular, he was eager to get "the locking shot": a shot of Olly and Suzi with one of the leopard seals. Describing the limited visibility in which a seal's presence only became apparent when it was already very close, Olly said:

> We took some chances on a few occasions, in deep water with these leps. I remember hanging on, I put my hand into an ice hole, gripped on, and Suzi was on my right, and my buoyancy was all over the place because we were in this swell, and the lep poked his head around this iceberg and he just came around within inches of us and it was like—whooo—we're holding the

drawing board and trying to draw it, and we know we've got Doug on the back of our shoulder, trying to get it all right, they call it the locking shot.

Pushed further on the question of how, in the face of the distractions of extreme temperatures, risk, physical discomfort, and lack of dexterity, they nevertheless managed to attend both to the animal and to the drawing, Suzi responded:

> The nice thing about that is that it's what we've been doing for so many years, and it's actually the easy bit, it's the bit that *I* like, because it takes my mind off everything else, like I hate this, and am I going to drown?, and all those fears in my head. If I'm concentrating on a drawing, that's what I'm really happy doing, so I'm completely relaxed in that sense, and while this animal's whizzing around I'm just trying to get as much information down as possible. That actually is the easy bit.

The gathering of the information is the key thing. Olly described his elation at their encounter with the looming seal, regardless of the difficult and rather frightening circumstances: "I remember just thinking to myself, great, now we can do a thousand paintings about leopard seals, it really doesn't matter, because *we've seen it, there.*" The issues raised in this whole exchange say a great deal about the artists' attitudes and values and therefore perhaps deserve to be contextualized more fully.

Attentiveness and Creativity

The relation of attentiveness and creativity (including attentiveness to the more-than-human world) has recently been explored in Wendy Wheeler's important 2006 book *The Whole Creature*. She contends that creativity "lies in being able to 'disattend from' the logic and rules of grammar" and of other rule-bound forms of meaning making or understanding, "in order to 'attend to' the perfusion, and disorderly profusion, of many signs which aren't supposed to count as legitimate" if understanding is thought about "only in linguistic rule-bound terms." She continues: "This 'attending to' is both conscious and unconscious—a kind of free-floating attentiveness . . . which poets have long described as waiting for the muse to descend" (146).

In one sense, this delicate balance—of disattending from rules in order to attend to that which those particular rules render illegitimate or irrelevant—does not quite match Olly and Suzi's account of their own attentiveness (both to the animal and to the drawing) in extreme conditions. But there are certainly forms of disattending from in order to attend to at the heart of their practice. At the most obvious level, it is their attention to the demands of the mark-making process that allows them to disattend from all of the distractions pressing in on them. At another level, like any artist with concerns for the environment and for nonhuman life, their thoroughgoing attention to endangered creatures and habitats is itself a committed disattending from Western culture's broadly anthropocentric and inward-looking value system. When Wheeler characterizes "a spirit of continual curiosity and attentiveness to what's going on" as "an openness to life" (147), she finds a phrase that could hardly be bettered as a description of Olly and Suzi's outlook.

Wheeler would almost certainly recognize and be sympathetic to another aspect of Olly and Suzi's description of how they work. It is that in the urgent moment of capturing something of the animal's presence, *while in its presence*, the distractions from which they must disattend are a vital part of the experience of attending. This is why their own accounts of their physical circumstances and states of mind seem so pertinent: not in order to privilege some notion of the "artists' intentions," nor to reduce their art to the story of its making, but rather because it seems clear that they are describing a bodily or an *embodied* attentiveness.

Building on Michael Polanyi's work in the 1960s on tacit knowledge, Wheeler sees the knowledge on which any form of creative practice draws not as abstract, rule-bound, and rationally acquired but rather as something more slowly and thoroughly gathered, learned, and embedded: she calls it "the tacit experiential knowledge which lives in the bodymind" (147). Drawing also on her wide reading in complexity theory, she suggests that organisms (including human beings) can only ever be understood in relation to their environments, and that this kind of thoroughgoing connectedness—whose ecological implications are clear enough—makes it entirely meaningful to speak of "body-mind-environment systems" (117). Olly and Suzi's repeated emphasis on the importance of their firsthand experience of the wild ("no experience, no art") and their descriptions of how their bodies dealt with that experience seem closely to echo Wheeler's notion of a creative, skillful, and generously inclusive reaching out for new knowledge and new experience: "We incorporate it in our body—or extend our body to include it—so that we come to dwell in it," she writes (63).

Accuracy and Beauty

Olly and Suzi frequently comment on the beauty of the animals they draw and paint, but asked whether that beauty had to be conveyed through making their own work beautiful—whether beauty had its own pictorial work to do, in other words—Suzi reflected: "That's a really good question, and I don't know the answer." Thinking about their work on sharks, off the coasts of both South Africa and Mexico, she expressed the view that "we should be trying to show people that they're not just big scary animals. They *are* beautiful, so it's my place to show other people that they are." Asked again whether this had to be achieved by making the image itself beautiful, she answered: "I think the image needs to be as *accurate* as it can possibly be." At this point Olly offered an important clarification: "But that's accurate to *our* experience. It's got to be as accurate as possible to our experience of being in five meters of choppy swell in a crappy cage with each other banging around and this thing trying to eat us. That's our interpretation of what's real and beautiful."

Three specific examples of this embedding of their own embodied experience into the image can be offered from the Antarctica trip. One of Duffield's photographs of the leopard seal subsequently won him a BBC Wildlife Photographer of the Year award in the underwater photography category. It was, however, another of his photographs of the same animal (Figure 7.2) that the artists and photographer had selected to count as part of Olly and Suzi's output because, Olly explained, that particular one summed up the "ominous vision that we best remember."

A second example concerns a drawing made on board the Pelagic. It was based on a number of drawings they had made "from life," face-to-face with creatures encountered over the preceding days (Figure 7.3). (In this sense, it resembles their new working method in the studio, where earlier images may be purposefully recombined.) The drawing is made up, Olly explained, of "images we'd been toying with previously": a crabeater seal, a pair of penguins, and a jellyfish that Suzi had seen and drawn underwater.

Here, as the word "toying" may itself suggest, notions of accuracy, beauty, and pleasure stand in a complex relation. Characterizing this kind of composition as something of "an experiment" for them, Suzi described her thought process: "I'm just drawing this yellow ribbony bit of jellyfish and thinking, well, I don't really know if it was in that position but I want it to go like that, so I'm using my imagination, having a great time, not worrying that it's actually accurate." Speaking of the editing of earlier drawings that led to this particular composition, she explained: "We chose three things that we thought were really important in Antarctica—important because we'd seen them." They both acknowledged that in the abstract, at least, it might have seemed

Figure 7.3. Olly and Suzi, *Cycle of Prey*. Caran Dache pencil on paper, Antarctica, 2005. Courtesy of the artists.

important to include an image of krill in such a composition because of its centrality to the life of the region. But there is only so much that can go into a successful drawing, and, Suzi remarked, "Visually we wanted to use our jellyfish because we love that image, and it's a beautiful thing to draw."

The third example concerns a simple linear image of a pair of seals, drawn on a map that the artists had been given by the skipper on the boat (Figure 7.4). As Olly pointed out, "It's the role of our drawing and painting to focus," and often all that is needed is a very "spare drawing": "Drawing is decision making; we can only do so much, we can only use so many colors, or we might just as well take a photograph. So to try to do too much with your drawing and painting would I think be a mistake." Suzi added: "But it's also nice, with this one on the map, just to draw the seals very simply down there in the corner, on the blank bit. That sort of says enough: *they're there, and that's how we're seeing it*. In the three weeks we were down there, we saw such a tiny part of what there is down there, and that's what we came back with."

Figure 7.4. Olly and Suzi, *Here Be Mermaids*. Acquarelle on old Antarctic chart, Antarctica, 2005. Courtesy of the artists.

Unpictured Habitats

The drawing on the map—which at least gives the seals a schematic geographical location—points to a rather unusual feature of the vast majority of Olly and Suzi's drawings and paintings. Given their very clear environmental concerns, it is striking that their images of animals are generally placed on the blank ground of the white or cream paper they typically use. The landscape is left out. In the drawing of penguins playfully skidding and sliding down a snowy incline (Figure 7.1), admittedly, the white ground almost inadvertently stands in for that little patch of Antarctic landscape, but that really is something of an exception.

Olly and Suzi made three points in relation to this observation. One was that always to draw the animal in its habitat would undermine the distinct role of the photographs that also form part of their body of work. The key thing is to use each medium to its best advantage:

to know what works. Their second point was that there is no shortage of more conventional wildlife art that places animals in the landscape, where "it's always the tiger in the bush or wolves tracking across the hills," as Olly put it. He wanted not so much to criticize the comforting associations of much of that popular art but rather to emphasize his and Suzi's singular focus on the animal: their attempt visually to sum up "the spirit that *we* took from the animal." "Not," he adds, "that any of these drawings define a species, not at all." They are simply the attempt convincingly to capture "a very brief second that we might have seen." But as Suzi frankly acknowledged, there also was the practical matter that whenever they *have* placed the animal in its landscape in a drawing, "it's never worked."

The third and most telling comment they offered concerned their own temporary inhabiting of these habitats. Suzi insisted: "We're going out to this environment, and we're in it, so that's our moment of it, it's us being there in it, and then specifically choosing what to put in and what to leave out."

"The Whole Chain"

Pressed as to why these animals matter, and more particularly as to what would be lost if they were no longer there, Suzi's simple response was "our world." Olly answered more fully:

> There'd be nothing to venture for, nothing to keep us on our toes. If you're walking through the Arctic tundra just picking daisies, not having to worry at the back of your mind about polar bears, you're left with a very severe habitat with not a lot going on. And it's got massive ramifications for us as humans, because then suddenly all it really is about is exploiting it: if there are no animals left, the lobbyists can't whine, so let's rape it, drill it, mine it, build on it, tarmac it. And then we've really lost something.

Quite apart from the issue of commercial exploitation, they both emphasized the ecological interdependence of species and their habitats. Olly offered a rather detailed hypothetical account of the environmental consequences if caribou were to disappear from the Brooks Range in Alaska, and Suzi used this as an opportunity to relate their environmental concerns to the priorities of their own image making: "You start understanding the whole chain, and why each thing is vital for the next thing, and that's why it's important not just to focus on one big scary creature. The whole essence of the place would be nothing without all those hundreds of little pieces of the jigsaw puzzle."

This last comment, about big scary creatures, certainly reflects Suzi's extreme frustration at the extent to which the 1997 "shark bite" photograph discussed earlier has continued to dominate public perception of their work. Dramatic and effective as that photograph of an animal interacting with its own painted image may have been, Olly and Suzi are wary of being expected to produce sensationalist imagery that may only reinforce popular misconceptions of the wild.

The artists' acute and growing awareness of "the whole chain" in the ecosystems they visit is, of course, no guarantee that their artworks will effectively contribute to the preservation of the links in any of those chains. Indeed, contrary to the views of those who "equate knowledge with control," Thomas A. Sebeok—writing from the perspective of complexity theory, which has influenced much recent ecological and eco-critical thought—contends: "The *more* we understand the complexities of a system, the *less* we should be confident of our power to manage it" (qtd. in Wheeler 2006, 121–22). But as Wheeler points out, "With that strange forward directedness of life itself" that might be said to characterize any creative outlook on the world, artists rightly go about their work with "a general confidence" (147).

That confidence in embodied skills is what Francisco Varela calls the intelligence that guides action, but that does so, crucially, "in harmony with the texture of the situation at hand, not in accordance with a set of rules or procedures" (1999, 31). This seems an apt enough description of Olly and Suzi's approach, as does Varela's account of a creative outlook as "a journey of *experience* and *learning*, not . . . a mere intellectual puzzle that one solves" (33). Readers and viewers will draw their own conclusions about the effectiveness of the images and strategies employed in Olly and Suzi's work, but even to ask a question along the lines of "Does their work *work?*" may be to ask the wrong kind of question, because it is simply too closed a question. Better, perhaps, to return to Marvin's continuum of humans seeing, looking, watching, and observing animals, and to think about where a creative attending-while-disattending—the manner in which artists might look at animals, in other words—would fit on that continuum. In a sense, of course, it does not fit; it is more like an interruption of it. And this is the modest claim that can be made for art such as Olly and Suzi's: that once in a while, at least, and without the artists necessarily even recognizing that it has succeeded in doing so, this art puts into place a creative interruption of the ways in which humans habitually look at animals.

Notes

1. Olly and Suzi, unpublished interview with the author, London, August 22, 2005. All further unattributed quotations from the artists are drawn from this interview.

2. The film, directed by Rupert Murray, is entitled *Olly and Suzi* and is an Arcane Pictures production.

Works Cited

Baker, Steve. 2000. *The Postmodern Animal*. London: Reaktion Books.

Berger, John. 1980. "Why Look at Animals?" In *About Looking*, ed. J. Berger, 1–26. London: Writers and Readers.

Dobrin, Sidney I., and Sean Morey. "Ecosee: A First Glimpse." Introduction in this book.

Marvin, Garry. "Guest Editor's Introduction: Seeing, Looking, Watching, Observing Nonhuman Animals." *Society & Animals* 13:1 (2005): 1–11 (Special theme issue: Ways of Seeing Animals).

Morey, Sean. "A Rhetorical Look at Ecosee." Chapter 1 in this book.

Olly and Suzi. 2003. *Olly & Suzi: Arctic, Desert, Ocean, Jungle*. New York: Abrams.

Polanyi, Michael. 1967. *The Tacit Dimension*. London: Routledge & Kegan Paul.

Varela, Francisco J. 1999. *Ethical Know-How: Action, Wisdom, and Cognition*. Stanford, CA: Stanford University Press.

Wheeler, Wendy. 2006. *The Whole Creature: Complexity, Biosemiotics, and the Evolution of Culture*. London: Lawrence & Wishart.

Wolfe, Cary. "From *Dead Meat* to Glow-in-the-Dark Bunnies: Seeing 'the Animal Question' in Contemporary Art." Chapter 6 in this book.

8

Connecting with Animals

The Aquarium and the Dreamer Fish

Eleanor Morgan

In 1998, a group of fishermen working off the coast of California caught a small black fish that had a thin rod attached to its head. An oceanographer onboard the vessel identified it as *Oneirodes acanthias*, or dreamer fish, a member of the anglerfish genus that can live at depths of up to 6,000 feet. The female dreamer fish has a bioluminescent lure protruding from her head, which she manipulates to attract mates and prey. The males, tiny fish with huge nostrils, are one tenth or even one hundredth the size of the female. Attracted by the female's light and smell, the male attaches himself by biting into her body. The bloodstreams of the two then fuse, allowing the male to live parasitically on the female for the rest of her life. This fusion creates a new unit, a hermaphrodite that is able to produce both eggs and sperm (Pietsch 2005). The particular specimen that was caught in the trawler's nets was female and had a huge swollen belly. The stomach was dissected to reveal a half-digested ball of packing tape. The fish had apparently mistaken the tape for food and swallowed it whole (Sendall 2004).

I recently have had two different encounters with local animals: the first with sea creatures at the Vancouver Aquarium and the second with the preserved dreamer fish specimen, which is now shelved in the fish collection at the Royal British Columbia Museum. Both experiences were fishy, but while the aquarium allowed me to see an underwater world full of living creatures, the dreamer fish is so physically adapted to the deep sea that it would never survive in an aquarium and is most commonly seen preserved in a jar (in this case, in a jar labeled 998-323-1, complete with packing tape). A different form of preservation is at work in the aquarium, in which individual living animals are

169

studied and bred to aid the survival of species whose disappearance is a real and an imminent threat. Dr. Georgina Mace, director of science at the Institute of Zoology in London and member of the Millennium Ecosystem Assessment team, states, "As we move into the next 100 years, we'll be seeing extinction rates that are a thousand to 10,000 times those in the fossil record" (Amos 2005). This acceleration points to the effects of human production and consumption on the environment: everything from invisible carbon emissions to a roll of packing tape discarded into the ocean.

Our own existence is intimately tied to animals, linked as we are through a shared planet and strings of DNA. Within this sea of connections, I focus on my two animal encounters. In the first, I looked through the glass of an aquarium; in the second, I looked through the glass of a jar. I discuss how each experience offered me a different understanding of our connection with animals, and how their existence and fate are perceived through our own.

Figure 8.1. The Vancouver Aquarium. (author's photo)

The Vancouver Aquarium is an animal research facility and popular attraction in the center of Vancouver. It houses an array of local marine life, in addition to creatures from farther afield, such as Arctic beluga whales and tropical fish. Visitors can watch the larger fish and mammals feeding or performing various tricks in response to their trainers' orders, nodding their heads or leaping into the air. One of the most popular attractions is a pair of sea otters, often seen lying on their backs holding hands. In the wild, this has practical uses, as it prevents them separating and drifting apart at sea. In the aquarium, it makes for a particularly adorable scene. Like other contemporary animal facilities, the Vancouver Aquarium attempts to combine conservation and scientific research with public appeal and entertainment. Visitors supply some of the much-needed research funds, and it is hoped that in return they leave with a greater understanding of marine life and the threats to the underwater habitat.

As I queued to buy my ticket, I noticed these words written above the entrance to the aquarium: "To inspire, to connect, to conserve," accompanied by the phrase, "It's another world." These lines suggest that the purpose of the aquarium is both to protect species and to offer the possibility of connecting with an alien world. As a visitor, I am encouraged to take on the job of animal protector—to learn from what I see in the aquarium and apply it to life "outside." The aim of connection is essential to this scheme. If I connect with animals, it implies, then I will be more inclined to protect animals. The nature of this connection becomes clear from another line written on the entrance wall: "For in the end we will conserve only what we love, we will love only what we understand, and we will understand only what we have been taught." This quotation, from the writing of Senegalese conservationist Baba Dioum, appears in public relations material in zoos and aquariums throughout the world,[1] and it appeals to an emotional and a loving connection achieved through understanding. It is an idea based on the hope that we will protect the ones we love. The role of visitors to the aquarium is to look at animals, to connect with animals (with love), and to conserve animals. The natural environment is disappearing, and the aquarium offers us a chance to look at and connect with a dying world.

To encourage connection through the bars of the cage, aquariums and zoos often emphasize our biological link with nature as Homo sapiens. An extreme example of this was attempted in 2005 at the London Zoo. Organizers caged eight human volunteers in the sloth bear display for one week, all of them naked except for some carefully placed fig leaves (the bear had been removed from the exhibit). This was intended to highlight our own position as human animals and therefore to encourage us to protect our animal relations. However, we have become so used to observing people and being observed that the experiment was more reminiscent of a television reality show. Rather than raising awareness that people are animals too, the experiment encouraged the idea that keeping animals in cages is entirely "natural," simply an extension of human entertainment and society. Conversely, it naturalized the practice of observing humans by equating it with the observation of animals.

We define society in relation to our understanding of nature. In this comparison, "nature" is a constant over which we have little control, while "society" is a changeable state. Historically, this division has allowed Western thought to use nature as an unwavering symbolic site on which to project its idea of humanity. Thus we invoke nature as something outside of human control, while paradoxically we alter our definition of what is natural in accordance with changes in society. In *Politics of Nature*, Bruno Latour writes:

> Not a single line has been written—at least in the Western tradition—in which the terms "nature," "natural order," "natural law," "natural right," "inflexible causality," or "imprescriptible laws" have not been followed . . . by an affirmation concerning the way to reform public life. (2004, 28)

When we try to connect with animals, we are attempting to leap across this nature/society divide. Animals are at once both like and unlike us. We are part of nature as Homo sapiens, yet we remain distinct through our societies, our cultures, and our languages. We embrace this awkward connection through anthropomorphism, in which we understand animals metaphorically through our own behavior; thus, lions are "brave," and foxes are "cunning." The creation of language, the capacity for symbolic thought, is one of our main distinctions from animals. John Berger argues that the animal metaphor is the origin of all language, our first attempts at symbolizing something outside of ourselves (1991, 7). Therefore, the animal metaphor ties us to other species while simultaneously emphasizing our separation from them. In his discussion of the animal metaphor, Christopher Tilley writes:

> [O]bservations of the characteristics of animals inform the metaphorical workings of the human mind, providing it with raw materials for processing and informing an understanding of human society. Such a self-understanding then further informs a conceptualisation of the world of animals. (1999, 50)

When we look at an animal, we are attempting to cross divisions between nature and society, animals and humans, essential to how we construct our perception of the world around us and ourselves.

At the Vancouver Aquarium I am able to stroll along carefully designated paths and to stop at certain points, with the promise that I will experience the thrill of watching live animals. I see a living collection, with each animal species, including my own, in its own distinct environment. The difference is that the animal's space is a theatrical stage, offering the minimum space and props required for the creature to survive. It could therefore be argued that the entire space is a human construction, our own imitation of nature. However, within that imitation we construct barriers to prevent access to the animals. It is thus important to emphasize that the aquarium is not a false construction of a real nature but symbolizes a real relationship between humans and animals that goes beyond the walls of the institution. It is a relationship in which we are spatially and temporally distanced from the animals that we observe and the nature that we are told to protect.

In the aquarium animals are removed from their natural environment and grouped together to create an unlikely geography, in which an Arctic beluga whale swims meters from a tropical fish display and the outdoor café. Within this collection, the animals are representatives of their species; they reference a world "elsewhere." At the aquarium, I am looking at living animals in a real setting (there is nothing imaginary about the water, the glass, and the café), yet the animals in need of protection exist "outside," in nature. Captive animals can only represent their species by being less real, less alive to us than their wild cousins. However, as symbols of nature, they alter our understanding of all creatures, captive and wild.

In her discussion of collections, Susan Stewart writes, "The collection replaces history with classification, with order beyond the realm of temporality" (1993, 151). Animals in the aquarium exist in a nontime. They are representatives of their species and as such can be replaced. There is therefore a sense of continuous existence, coupled with the lack of a living "being." Aquariums are a site of preservation not of living animals but of animals that will neither truly live nor die. The structure of aquariums, with artificial and real nature mixed together in displays, echoes the museum diorama.[2] Yet in this case, rather than the stuffed animal behind the glass hovering on the edge of life, the caged animal occupies a state of animated death.

Is it possible to see beyond the collection at the aquarium? What happens when an animal's gaze meets our own? Could this be a moment of connection, in which human and animal are joined at the eyes, both aware of looking and being looked at? At the aquarium, the distance from viewer to animal is not too close to be dangerous and not so remote that the animal is a dot on the landscape. The distance between viewer and animal mirrors the distance between hunter and animal, and in this look we are forced to play out the moment before capture. In *Savages and Beasts*, Nigel Rothfels writes:

> We seem intent to catch the look of an animal, to see the animal look at us. This is, of course, the look that so many hunters talk about, the look caught in taxidermy the world over. (2002, 32)

At the aquarium, we long for the animal to look back—a yearning for some contact with a disappearing nature. Yet our moment of desired connection throws us into the "wrong" historical moment—a memory of the human extinguishing the animal. Eye contact does not reveal the animal; it reveals the constructions that keep the animal in its place. Therefore, the moment we believe there will be the most connection is instead the moment we realize our absolute separation.

Rather than being specific to aquariums, this situation is part of a gradual shift in animal visibility that began in Europe and North America during the Industrial Revolution.[3] Prior to this, animals shared humans' space as meat, companions, curiosities, sorcerers, and laborers. They were not restricted to one category but could exist in several of these states. Therefore, the same pig could be both companion and meal (Berger 1991, 7). In the nineteenth century, machines began to replace laboring animals, and the visibility of animals was restricted; the living animal was separated from the meat on the plate. With this development came an increase in the popularity of household pets. A pet was certainly not a companion to be eaten or used for labor, but was, like any personal property, an extension of the owner's identity. This categorization of domestic animals coincided with a Western craze for seeking out foreign "exotic" animals to be brought home and observed either stuffed in a museum or caged in a zoo.[4] These were the animals to be consumed *visually*. Understanding of animals was therefore compartmentalized into those to be eaten, those to be owned, and those to be looked at. Increasingly, it became socially unacceptable or absurd for these categories to be broken, hence, we would not keep a cow as a pet, or eat an elephant, or keep Labrador retriever dogs in zoos.[5]

While the everyday visibility of real animals has decreased, the marketing of animal representations has become prolific. The stuffed toy animal is a commonplace product, safer and easier to care for than a pet, always for sale in the zoo or aquarium shop where children can take home the cuddly version of the live animal. Zoos and aquariums encourage an emotional connection with animals that echoes a child's love for a cuddly toy. In this relationship, the object of affection is a physical extension of self-love. The child loves the toy, which is physically distinct but without the independent willfulness of another child. The child can then imagine the dumb toy to reflect his or her love. A similar relationship exists between visitors and animals at the aquarium. This is not simple anthropomorphism. It is dependent on a complete denial of animal difference, or separate existence. This act of humanizing animals is a common contemporary method often used to promote environmental concerns. For example, the animated film *Happy Feet* (2006) depicts a group of singing emperor penguins whose survival is threatened by overfishing. It is up to Mumble, a tone-deaf, tap-dancing outcast, to save them. After a journey of discovery (and self-discovery), Mumble alerts humans to the penguins' predicament by organizing a synchronized dance routine, which finally leads to a change in global fishing laws. *Happy Feet* is an entertaining cartoon that, like the aquarium, draws on a human emotional connection to heighten environmental awareness. However, this can only be achieved through removing animal distinctness. Animals as real, living things are replaced by intangible substitutes: humans disguised as cartoon penguins. In this masquerade, animals become another product for us to consume.

We have deliberately created a situation in which we physically distance ourselves from animals while encouraging a desire to connect. The purpose of the aquarium is to protect animals and support that protection in the wild. However, within such spaces animals unavoidably become little more than consumer objects, symbolizing a far-off nature. The animals at the aquarium are both real and unreal, and their real disappearance has a similarly unreal quality.

How can we conserve something that is not real or alive to us?

Consider now my second animal encounter, with the dreamer fish in the Royal British Columbia Museum. This fish is one of thousands of specimens housed in a huge room with high ceilings, filled with row upon row of fish in jars. The flesh of the dreamer still has the same firm, squashy resistance of any raw fish; the physical sensation of touching it is uncomfortably reminiscent of holding a store-bought salmon. Visually, however, the dreamer fish is very unappetizing; it has large sharp teeth, black wrinkled skin, and a thin lure coming out of its head. I would not want it in my mouth.

Now preserved in a jar with the packing tape that killed it, the dreamer fish has been removed from its natural habitat, the alien space of the deep ocean where humans rarely venture. Its fantastical appearance evokes mythical tales of deep-sea monsters or brings to mind the story of the donkey with a carrot and stick (the fish with its own rod). Then there is the wonderfully romantic name, the "dreamer," floating through black waters following a light, an image suggestive of endless hopefulness.

I decided to create a wax replica of the dreamer fish as part of an exhibition. The process of making the sculpture was similar to the method used in the nineteenth century to create wax models of human anatomy, which combined scientific knowledge with artistic techniques. This practice had died out by the turn of the century, along with most of the technical knowledge. I was therefore obliged to experiment with casting and pouring, and I finally created different shaped tools from bits of metal, which I heated in a flame and used to simultaneously carve and melt the wax. The most difficult part of the specimen to reproduce was the packing tape. I bought a roll to study. It was a very tough material, reinforced with thin lines of stringy plastic. Nineteenth-century wax sculptors had used string dipped in wax to form human veins and arteries, and I decided to attempt something similar for the lines in the packing tape by embedding dental floss into the wax. I tried to tear off the roll a strip of packing tape with my teeth, and my throat gagged in response. The tape is impossible to bite into—it is solid, "plasticky," sticky. Once the dreamer fish had mistaken the packing tape for food, she could not stop until she had swallowed the entire thing.

Figure 8.2. Digested packing tape found in the stomach of the dreamer fish (author's photo)

My experience of creating a replica of the dreamer fish included a combination of scientific and artistic experimentation and a chain of things from various histories and places linked by association to the dreamer fish. These "things" range from the physical (the jar in which the dreamer fish was preserved), to the mythical (images of deep-sea monsters). I propose that following this process of experimentation and "things" offers an alternative to the emotional connection with animals promised by the aquarium. Although it was an individual experience, it enabled me to glimpse the reality of the dreamer fish.

Experimentation

Science dominates our understanding of material reality—of the factual nature of things (Tiffany 2004, 75). These material facts are the result of a process of actively observing and experimenting on the world around us. There is therefore a level of uncertainty and curiosity to an experiment; it is carried out with the hope and possibility that something unknown might be discovered. In the aquarium, I can look at animals through glass and learn about their behavior and habitat from information panels. Yet the experimental process that has gone into constructing the material "fact" of animals is not accessible to the visitor, and with it is removed the imaginative potential of curiosity. In his discussion of scientific realism, Ian Hacking writes, "New ways of seeing, acquired after infancy, involve learning by doing, not just passive looking" (1983, 189). If experimentation is an active process of determining reality, then I must adopt some of the techniques of the scientist if I am to understand the dreamer fish as real.

How can this reality be determined? Hacking argues that making representations shapes our concept of reality. It is through creating likenesses that we are able to distinguish one from the other, and it is only then that the concept of reality is formed (Hacking 1983, 136). From this argument it follows that by creating a representation of the dreamer fish, rather than looking at it passively, I understand the fish as real. The final object does not offer this reality, but it may be found through the active process of making that object. This process of representation comprises actively experimenting on the world and observing the results—a method of determining reality through cause and effect.

As an uncertain process, an experiment is not the exclusive domain of the scientist but is essential to any form of representation. It recognizes and creates relationships between active participants, including people, objects, and ideas, which may include all kinds of illogical or illegitimate associations. However, if I include *everything* in my experiment, then I am in danger of getting swept along in a theory of ever-expanding networks. For the experiment to be effective, there must be some attempt to record its results (Latour 2004, 200). I must use and create a dynamic narrative from the uncertain experiment that details my relationship to the fish. In this way, my understanding of the animal is not reliant on fixing it *visually* but implicates my own shifting narrative and history. It is animal as activity rather than animal as object. What would this narrative look like? As a possible answer, I turn to the chain of "things" that I attached to the dreamer fish.

Things

My experiment with the dreamer fish resulted in the association of diverse "things" from different places and times and in various forms: inanimate materials (packing tape, dental floss, string, wax); dissected animals (fish, human anatomy); sites (museum, specimen jar, art studio); and mythical images (donkey with a carrot and stick, deep-sea monsters). All of these things contribute to my understanding and construction of the dreamer fish. In this sense, I can only understand the dreamer fish as a living thing through other things. Sometimes these combinations of things and their effects can be surprising, such as my physical reaction to the packing tape. This was a nonvisual momentary experience that did not fix the dreamer as an object but offered a partial glimpse of the dreamer as real through the very thing that killed it.

Our understanding of animals is caught up in these systems of science, myth, and material, and yet we still attempt a distinction between the nature that we must preserve and the society in which we live. This divides our own fate from that of animals, although we are biologically and geographically linked. Because of this connection, I cannot consider the dreamer fish simply as an object to be looked at but as a spreading thing that sticks to me. The etymology of the word "thing" suggests this approach. In many European languages, there is a link between the word "thing" and ancient words for an assembly or a gathering, such as the Icelandic term for Parliament, *Althing*. Therefore, a "thing" also may refer to a collective in which issues are communicated and discussed. An animal as an object to be looked at is fixed and inaccessible. An animal as a thing suggests that we are implicated in its construction, and that it offers a route of attachment.

Using the thing as a dynamic form of connection means that we do not pretend that we can connect with animals directly, therefore denying their distinctness, but that we can connect through a chain of things. There is no constant nature, no lost paradise, but an associative collection of things. To follow this will not result in a final object but will offer a glimpse of the material "thinginess" of an animal. In his discussion of "Thing theory" Bill Brown writes:

> We begin to confront the thinginess of objects when they stop working for us . . . when their flow within the circuits of production and distribution, consumption and exhibition has been arrested, however momentarily. (2004, 4)

These "circuits" include the effects of human production and consumption, the very systems that are destroying animal species and habitats. By following these circuits, I am linked directly to the animal and its disappearance. The "thinginess" of the dreamer fish was for a moment apparent to me when I gagged on the packing tape that killed it. In this second, the fish escaped its place in the circuits of production and became not an object to be visually consumed but a distinct and real thing whose fate is linked to my own.

In conclusion, it is not a question of whether or not connection is desirable, but that it is absolutely unavoidable. We are already connected to animals—it is just that the connection is not as attractive as the love we feel for a stuffed toy or pet. Our connection to animals includes unappealing things such as avian flu, the malaria virus, and the waste of production. The connections keep on multiplying. In attempting to conserve animals, we must understand them as real living things through the very system that is destroying them. Until we understand animals through this connection, they are both already lost to us (as unreachable) and indestructible (as preserved). We are linked to animals through mortality, yet an animal in the aquarium never really dies—there is always another of its species to replace it. If we only understand animals through these "viewing spaces" then their mortality has no reality for us.

In dealing with our connection to animals, I have left one fundamental question unanswered: Why conserve animals? The answers are numerous and vague. There is the aesthetic argument, that without the wonder of animals the world would be a poorer place. There is the ethical argument that, like us, animals have a right to exist; and there is the more anthropocentric argument, that their survival is tied to our own. My own answer, in addition to all of these reasons, is that a world in which we are the only species would be too unbearably lonely to imagine. The disappearance of animals is not clear in the aquarium. We must understand animals through the real messy space that we share with them—a space that exists between nature and society.

Notes

1. The zoos of San Francisco, Louisville, San Diego, Indianapolis, Frankfurt, Toronto, and Bermuda include Dioum's quotation in their mission statement or publicity material.

2. In "The Metaphoric Architecture of the Diorama," Stephen Parcell writes that in the diorama the animals' "vivid poses and background settings suddenly resurrected these creatures into an intermediate state somewhere between life and death" (1996, 181).

3. On the categorization of animals in the nineteenth century, see Harriet Ritvo, *The Animal Estate: The English and Other Creatures in the Victorian Age*. (Cambridge, MA: Harvard University Press, 1987).

4. On the history of the establishment of the zoo, see Eric Baratay and Elizabeth Hardouin-Fugier, *Zoo: A History of Zoological Gardens in the West*. (London: Reaktion Books, 2002).

5. During the establishment of the zoological garden, it was not uncommon for scientists to eat the animals as an extension of their observational investigations. An example of this is a story about Frank Buckland, the early London zoo naturalist: "Buckland . . . was always 'in at the death' of any unusual mammal or bird, and ready and eager to dissect it and to enlarge his experience by sampling its flesh; with this in view he had once arranged for the exhumation of a leopard which had inconsiderately chosen to die while he was out of London, and the fire in the giraffe house in 1866 afforded him an even rarer and already roasted delicacy" (Blunt 1976, 201).

Works Cited

Amos, Jonathan. 2005. "Earth's Species Feel the Squeeze." *BBC News*. May 21 http://www.news.bbc.co.uk/1/hi/sci/tech/4563499.stm, accessed Aug. 15, 2005.

Baratay, Eric and Elizabeth Hardouin-Fugier. *Zoo: A History of Zoological Gardens in the West*. London: Reaktion Books, 2002.

Berger, John. 1991. *About Looking*. New York: Vintage.

Blunt, Wilfred. 1976. *The Art in the Park: The Zoo in the Nineteenth Century*. London: Hamish Hamilton.

Brown, Bill. 2004. "Thing Theory." In *Things*, ed. Bill Brown, 1–16. Chicago, IL: University of Chicago Press.

Hacking, Ian. 1983. *Representing and Intervening: Introductory Topics in the Philosophy of Natural Science*. Cambridge: Cambridge University Press.

Happy Feet. 2006. Dir. George Miller. Warner Bros.

Latour, Bruno. 2004. *Politics of Nature: How to Bring the Sciences Into Democracy*. Cambridge, MA: Harvard University Press.

Parcell, Stephen. 1996. "The Metaphoric Architecture of the Diorama." In *Chora 2: Intervals in the Philosophy of Architecture*, ed. Alberto Perez-Gomez and Stephen Parcell, 180–216. Montreal: McGill-Queen's University Press.

Pietsch, Ted. 2005. Personal interview. "The Dreamer Fish and Its Genus: Ted Pietsch, Curator of Fish, University of Washington." Artist's video.

Ritvo, Harriet. 1987. *The Animal Estate: The English and Other Creatures in the Victorian Age.* Cambridge, MA: Harvard University Press.

Rothfels, Nigel. 2002. *Savages and Beasts; The Birth of the Modern Zoo.* Baltimore, MD and London: Johns Hopkins University Press.

Sendall, Kelly. 2003. "Natural History: Dreamers." *Royal British Columbia Museum.* http://www.royalbcmuseum.bc.ca/Natural_History/Fish. aspx?id=294, accessed 10 April 2007.

Stewart, Susan. 1993. *On Longing: Narratives of the Miniature, the Gigantic, the Souvenir, and the Collection.* Durham, NC and London: Duke University Press.

Tiffany, Daniel. 2004. "Lyric Substance." In *Things*, ed. Bill Brown, 72–98. Chicago, IL: University of Chicago Press.

Tilley, Christopher. 1999. *Metaphor and Material Culture.* Oxford: Blackwell.

Part 3

Seeing Landscapes and Seascapes

9

Farming on Irish Film

An Ecological Reading

Pat Brereton

Until the 1960s, Ireland was culturally defined and economically determined as a rural, agricultural-based society. This preoccupation with rural values was augmented by a fixation with nationalism, a consequence of Ireland's long struggle to regain sovereignty and ownership of the land. In this broad-based revolutionary project, the landscape was appropriated by romantic nationalists to affirm its unique beauty and as a bulwark in the cultural and political struggle for independence. Now that the economy has been transformed through a "Celtic tiger" renaissance, there has been an apparent rejection of rural values alongside a form of amnesia about the traumas of the past. This is evidenced by a recent preoccupation with contemporary, urban-based generic cinema in Ireland.

Ecology has become a new, all-inclusive, yet often contradictory, metanarrative, which I argue is clearly present within mainstream film since the 1950s. This chapter applies some of the strategies used in my earlier work, *Hollywood Utopia* (2005), to provide a close reading of the narrative trajectory of *The Field*, *The Secret of Roan Inish*, and *How Harry Became a Tree* to explicate some of these broadly ecological ideas, while initially drawing on foundational classics, namely, *Man of Aran* and *The Quiet Man* to set up the study.

Farming as a primal rural profession, which oscillates between nurturing and exploiting the land, can be used as a barometer of core ecological land ethics. Much of the island of Ireland is mountainous and boggy and consequently of limited use for intensive farming. While not worth cultivating, such land nevertheless provides rough grazing and closely corresponds to the tourist industry's perception of "wild nature," ensuring the land's intangible worth for the economy.[1] Classics such as *The Quiet Man* incorporate the more staid, cultivated beauty of the landscape, while *The Field*, as we shall see, can be read as a contemporary cautionary tale on good land husbandry. This chapter will review a number of Irish films that foreground farming and, by extension, ecology. Like many dystopic narratives,[2] several cautionary stories on farming show the important role that film can play in promoting ecological awareness through the visualization of nature and the validation of a nurturing attitude toward the environment, thus highlighting the therapeutic force of popular film.

Nature as Form

A central tenet of ecology could be reduced to some form of "harmony with nature," along with the recognition of "finite resources." Everything else in this view is therefore either peripheral to, or at best ancillary to, these all-inclusive affirmations. An ecological imperative exemplified by organic farming seeks to reiterate a dominant global and holistic ethic for all sentient beings on the planet. In his *Dictionary of Green Terms*, John Button defines ecology and the growth of eco-politics as

> a set of beliefs and a concomitant lifestyle that stress the importance of respect for the earth and all its inhabitants, using only what resources are necessary and appropriate, acknowledging the rights of all forms of life and recognising that all that exists is part of one interconnected whole.[3]

Aldo Leopold was possibly the first to articulate this green ethic by declaring "a thing is right when it tends to preserve the integrity, stability and beauty of the biotic community. It is wrong when it tends otherwise."[4] E. O. Wilson, and his controversial theory of "biophilia" (1984), insinuates that because we evolved from nature, we still carry a part of nature in our hearts, and this is where humans feel their relationship with and responsibilities to the land, through the complexity of human-land relations. While Leopold recognized that humans have "inherently psychological affinities to the natural world, including aesthetic appreciation, emotional attachment, spirituality, and all of these affinities have evolutionary and developmental significance."[5] This chapter loosely applies Stephen R. Kellert's broad typology of this phenomenon—aesthetic, dominionistic, humanistic, naturalistic, moralistic, negativistic, scientific, symbolic, and utilitarian—to assess the ecological credentials of the representations of Irish farmers under discussion and to demonstrate when and how they are out of kilter within their rural landscapes.[6]

There is often little agreement among the large rainbow of green supporters on the specific means, especially the priorities and times-cales, for achieving harmony with nature. Simply looking for hope through an artificial development of holistic systems can be read as a recipe for disaster. Consequently, there is an inherent danger of simply endorsing the trend of using the therapeutic romantic representation of nature, including farming, to help audiences overcome the distresses of modern living.

Surprisingly, unlike the Hollywood evocation of nature, it would appear that representations of farming, at least in Ireland, signal a more dystopic vision of the land. As cultural critics suggest, our representation of nature usually reveals as much, if not more, about our inner fears and desires than about the environment.[7] Nevertheless, the two attributes can be regarded as coterminous, since our inner fears and desires often reflect or at least constitute in large part the "external" environment. For Henry David Thoreau, and later for Leopold, wilderness was a state of mind as much as a description of a place. Both men championed a land alongside a people ethic and not a land versus people ethic. Humans are a central environmental force shaping landscapes everywhere. Wilderness is our touchstone, "a question of intellectual humility . . . (giving) definition and meaning to the human enterprise."[8]

The Irish landscape in particular has been symbolically used in film to "facilitate violence, particularly in the use of cliffs and deep wells, as signs of abyss."[9] John Hill has illustrated the links between the apparent atavistic nature of the Irish Republican Army (IRA) and terrorist violence generally and their close representational association with raw nature on film. Most notably, the cliff tops, with their sheer drop into a rocky, grey, unwelcoming Irish sea, represent the last vestiges of control that the world, both natural and modern, can exercise over the lives of its inhabitants. The outline landscape is therefore imbued with a Godlike status, since the sea may not only "giveth but can also taketh away." This power of nature over its inhabitants became a dominant trope in Revivalist Irish literature, most notably exemplified in J. M. Synge's *Riders to the Sea* and its cinematic equivalent, Robert Flaherty's *Man of Aran,* which in turn serves to frame our understanding of the symbolic potency of the sea as dramatized in the Sheridan adaptation of J. B. Keane's *The Field.* There is surprisingly little cultural or literary analysis afforded to the agency of farming, except as a romantic abstraction. Nonetheless, it should be noted that at least one Irish poet stands out for his valorization of farming as a real physical activity, namely, Patrick Kavanagh.[10]

Luke Gibbons strongly contends that due to both its "colonial history and its position on the Celtic periphery of Europe, representations of Ireland over the centuries have been enclosed within a circuit of myth and romanticism."[11] The dominant myth visualized in Irish cultural narratives is a pastoral one that foregrounds an almost Arcadian evocation of the happy swain close to nature, alongside the cyclical rhythms of Earth. This light ecological myth was one fostered and encouraged by the cultural nationalists from the 1920s onward—most notably, by long-term political leader Eamon de Valera, who spoke of the Ireland of his dreams, where rural people lived the life God intended them by being at one with nature.

Man of Aran (1934) and The Quiet Man (1952)

This myth of the aesthetically pure Irish Westerner, striving to be "at one with nature," is affirmed in many seminal Irish films, from Robert Flaherty's famous "poetic documentary" Man of Aran to John Sayles's more recent The Secret of Roan Inish. All of these films apparently promote Kellert's aesthetic, moralistic, and symbolic typologies of nature. This relationship often is visualized by characters framed against a majestic landscape as they look out in awe and wonder. The static image of the father in Man of Aran breaking rocks with a pickax, framed against a majestic landscape and drawing up seaweed to fertilize his newly created land, remains a primal image for an Irish postcolonial consciousness. Their pride is symbolized by a frieze of an individual or a group of main characters creating a statuesque pose and counterpointed by the attention-seeking chords of celebratory Irish music, which dramatizes their mythic harmony with and in nature. Such cinematic juxtaposition recalls the generic format of the Western and road movies that have dominated many Irish films.

Man of Aran, filmed on the Aran Islands off the west coast of Ireland, ostensibly aspires to Kellert's category of scientific and documentary naturalism that remains a defining image of primitive cultural endurance and the romanticization of rural Irish identity. When the film was first seen by the political elite, it was said that Eamon de Valara—the leader of the Fianna Fail political party, who above all others helped define a postindependence identity that continued well into the 1950s—wept at the viewing with its affirmative portrayal of a heroic people. The film reflected a preoccupation with the West as defining a pure strand of Irish identity and was marketed internationally as an expressive document of stoic endurance and life during that period, foreshadowing our discussions of The Field.

John Ford's classic The Quiet Man, on the other hand, has remained the most enduring representation of a romantic Ireland, which has continued to appeal especially to a diasporic audience that enjoys its timeless, quaint, stereotypical, wild, and pleasure-seeking Irish characterization. Yet its technicolor vision of a land of "rosy cheeked colleens, leprechaun-like intercessionaries and humane clergy united in song, drink and public brawling, has little in common with the Ireland of the 1920s in which it is ostensibly shot."[12]

An authentic, if a nostalgic, Irishness has become closely associated with a rural landscape. As Gibbons suggests, Irish culture has a preference for a romantic outlook and a form of "soft primitivism," unlike the "hard primitivism" of *Man of Aran*, which was replaced by a more realist aesthetic that became more noticeable as the dominance of the rural economy decreased. Like W. B. Yeats's critique of materialism, America is represented as being obsessed with "lousy money" and materialist ambition, which drives Sean Thornton (John Wayne) back to the idyllic Ireland of *The Quiet Man* and his ancestral home, described as "another name for heaven—Innisfree." Unlike the Yank in *The Field*, Thornton's "coming home" is validated by the local community, which affirms his right to buy back the cottage where he was born. However, the closest Thornton gets to farming the land is by sowing flower seeds rather than more useful and utilitarian vegetables, as his wife initially scolds him. This is all very different from displays of an economic imperative in *Man of Aran*, where the natives eek out a hard existence on land and sea while struggling with the elements. The only exhibition of farm labor is in a communal threshing scene, where Thornton creates a more interesting personal spectacle as a fight is started with the squire over his wife's dowry. The success of the film outside of Ireland is largely based on postcard scenes of a romantic countryside that appeal to an American image of a touristic Ireland. Similarly, the more whimsical and self-parodying *Finian's Rainbow* (1968), or the recent derivative *Waking Ned* (1998), also provides a relatively successful tourist and stereotypical image of the island with landscape used to register familiar markers of Irishness. A more dramatic foregrounding of nature as a brooding presence is evident most noticeably in *The Field*, but first let us examine a lighter ecological and romantic fable.

The Secret of Roan Inish (1994)

John Sayles, best known for issue—oriented films such as *City of Hope* (1991) and *Matewan* (1997), wrote and directed *The Secret of Roan Inish*, which was set in Ireland, essentially because of favorable tax breaks. Translated as the island of seals, the film recounts a distraught father who has lost his wife and child, sending back his surviving daughter to her grandparents' house in a rural region of Ireland to seek solace and a new beginning. The ten-year-old girl, Fiona (Jeni Courtney), becomes enchanted by the mystery of her homeland, along with the local folk-tales, and she persuades her grandfather to allow a visit to their past by journeying from the mainland to this western mythical place that she calls home. Based on an Irish folktale and journey reminiscent of *Into the West* (1992) and its magical horse, Tír na nÓg, *The Secret of Roan Inish* deals with the myth of the Selkie, a magical creature that is a seal by day and a human by night. The original fable on which the film is based was written in 1957 by Rosalie K Fry and is set circa 1946, after the war. In both films, the human soul is reincarnated as an animal to protect loved ones left behind. Such Irish folktales call upon a strong oral tradition, retold through generations of families. Poet W. B. Yeats, for instance, frequently wrote about such stories and spoke of water fairies capable of shape changing and even as symbols of tumultuous weather.

From the opening sequences, a very clear opposition is made between the beauty of the island and the West and the impurity of the East and the city. This is witnessed through the innocent eyes of the young girl, with the camera framed from below at her eye level as she searches for her father in the pub. The voice of the barmaid affirms that this is nowhere to raise a child who needs fresh air. As in many stories, the child functions in the narrative as a conduit to a lost past, through the sea and wild landscape.

The journey begins with her arriving at her grandparents' little house, where well-known Irish actor Mick Lally confirms that the city is nothing but noise and dirt, yet he goes on to warn that "love of the sea is a sickness." The girl's sojourn in this primitive but beautiful place is framed by a number of stories told by her grandfather and a cousin, who is said to be a bit strange. The first story narrates the awful primal story of the girl's little brother and how he came to be washed out to sea in his cradle. As the islanders are getting together all of their belongings to leave the island, they are attacked, like in Hitchcock's *The Birds* (1963), by everyday seagulls, meanwhile, as the men are distracted, the cradle is taken out to sea by the tide. As they try to rescue the child, the elements suddenly turn nasty, with strong rain and winds, and the sky dramatically darkens. It is suggested by the storyteller that a malevolent force of nature conspires against the farmers on the island because they have abandoned both their natural habitat and their stewardship responsibilities. The trajectory of the fable involves a working out of this imbalance in nature as the islanders realize their loss and finally, with the intervention of the little girl and the return of the "lost" boy, move back to their native island.

While Martin McLoone reads the film as valorizing essentially regressive ideologies, with the Irish landscape layered and coded as a search for the meaning of lost identity, the director and writer suggest that the film has specific cultural resonance, since Ireland is obsessed with "loss"—epitomized by *The Quiet Man*—which has helped confirm it as a site of nostalgia for an American diasporic audience.

How Harry Became a Tree (2002)

While this strange art house experiment, directed by Goran Paskaljevic, came much later than *The Field*, it helps map a transformation from the light fantasy form of eco-worship displayed in *The Secret of Roan Inish* to a deeper, if pathological, variety in *The Field*. The film explores similar themes around sexual politics and the "sins of the fathers," which are framed against a continuing trauma over the death of a child. The bulk of the narrative focuses on acts of revenge and their consequences. Transposing a Chinese fable to post-civil war, rural Ireland in 1924, the Serbian director creates a potent metaphor for the Bosnian conflict as he explores the individual roots of conflict through a man who opts for misery over happiness and hate over love, as the tagline for the film affirms. Colm Meaney gives a rich, tempestuous performance of a gruff farmer called Harry who lives in the misguided belief that a man is measured by his enemies.[13] This is somehow analogous to the way a tree chooses the strongest nutrients for its roots. Inevitably, Harry proves to be his own worst enemy.

In many ways the narrative in this film is less important than the evocative cinematography that effectively captures the extremely inclement winter weather on the mountains around Wicklow and somehow seems to echo various European and other international sites of tension, where the environment dramatizes the personal conflict.

While these various creative intentions cannot fully hold the movie together, it nevertheless works very well as a theatrically driven character drama, with Harry transposing his soiled fingernails and well-worn nurturing farm hands to end up literally "dancing with the sky." This is as far removed as one can get from the plastic fantasy of the television franchise *Star Trek*, where the actor established his reputation. Harry all but parodies the most disparaged persona of the eco-activist or tree hugger through his total identification and adoration of this natural phenomena and, much to the consternation of his son, he pontificates on how a tree can overcome all that is thrown at it. "You could throw one of your Bible floods and it would still be here." Harry only wants to be rooted by the solidity of nature, foreshadowing Harris's seminal role in *The Field*. Having caused his only son, Gus, to emigrate, and no longer having an enemy by which to define himself, all that Harry has left is his dream of becoming a tree and being symbolically, if not literally, at one with nature.

In an important narrative incident, Harry tries to get the better of a rival by destroying the cabbages, organically and lovingly nurtured by him and for which he was well known in the area. All of his life-giving instincts are sublimated as he consciously breaks the necks of the healthy cabbages growing in his field, simply to deny his enemy the fruits of his labor. When this unnatural strategy goes wrong, he cannot of course undo the forces of nature as the vegetables rot in his field. Going against a primary land ethic accelerates his mental decline, which needs to be allegorically resolved, at least in the closing sequence, as he literally goes back to nature. As Leopold affirms, "We abuse land because we regard it as a commodity belonging to us," while the ideal is to "see land as a community to which we belong," and only then can we "begin to use it with love and respect."[14] This ecological lesson is dramatized by primal stewards of the land, including Harry the cabbage grower, and later the Bull McCabe in *The Field*.

The Field (1990)

While *How Harry Became a Tree* displays a form of sacred and primal model of farming, full blown if more conventional psychotic agency can be witnessed in the more mainstream *The Field*, directed by Jim Sheridan. The film reworks the classic Western trope and the representation of farmers as primal nurturers of land, who sometimes selfishly wish to enclose the landscape, thereby flying in the face of an idealized, romanticized open range. Here the central farmer is so fixated with enclosure and ownership that it borders on the pathological and contrasts with Irish travelers, who, like Native American Indians, are often represented as nomadic eco-warriors, while also demonized within the Irish nationalist mythos.

The late John B. Keane (1928–2002) wrote the play, and Sheridan, in adapting it for film, made some changes for dramatic effect. For instance, in the film version, the action was changed from a violent farming feud of the 1960s to the 1930s, and it is set in the remote bogland around Connemara in County Galway. The film also replaced a returned emigrant from England, which featured in Keane's play, with one from America as the narrative antagonist, possibly to make the film more palatable commercially. Ray McAnally had originally signed a contract for the part of the Bull McCabe but then tragically died before production commenced. Eventually, Richard Harris, with his powerful, furrowed face and Learlike hair and long white beard, made the role his own.

The film opens with two men in picturesque silhouette, pushing a cart over mountainous fields and unceremoniously ridding the carcass of a donkey into deep water below, thereby polluting the environment. The Bull later tries to teach his son Tadgh (Sean Bean) that harming the donkey, who apparently strayed into their precious field was wrong—affirming a clear sense of ecological values. But this does not extend to concerns over pollution, directly contrasted in the followed scene with the same two men gathering seaweed[15] in pouches to carry back up the mountains—a clear visual reference to the "hard primitivism" of *Man of Aran*—so that they can fertilize and regenerate the precious soil in the eponymous ultra-green field, which is the focus of identification and contestation in the film. The well nicknamed Bull clearly emulates Harry in *How Harry Became a Tree* as well as the islanders featured in *Man of Aran*. The Bull intones how he dug rocks and made the field breathe life, while displaying his massive hands. In the first of his many rhetorical speeches, the Bull pontificates to his son "God made the world," [but] "seaweed made the field," affirming his clearly defined organic and utilitarian land ethic.

Surveying the landscape, the Bull allegorically illustrates his total identification albeit fixation with the field: "This is what we'd be without the land," as he symbolically blows dandelion seeds across the landscape. At one level this can be read as an effectively dramatic piece of business to illustrate his understanding of the land ethic while paradoxically dismissing any transient notions of human agency and destiny that promote aimless travel. Looking at the airborne seeds from a scientific perspective, however, far from being aimless, such travel embodies a natural and an evolutionary design feature of the flower, ensuring that suitable new sites are found well away from the parent plant to begin germination and a renewal cycle. At another level one could hypothesize incidentally that the freedom of effortless travel corresponds to the romantic dream of the nomad, typified in Ireland by the gypsy or traveler. The Bull's alter ego is embodied in the traveler (a term used to describe an Irish gypsy with a nomadic lifestyle), which he condemns out of hand as aimless and lacking utility. Such polarizing notions of place and space as definers of human identity are reminiscent of the Western and its representation of Native American Indians and their nomadic agency, which are at the same time read as more intrinsically ecological and at one with nature than more ownership fixated farmers, whose raison d'etre is to enclose land and maximize productivity.

Later the Bull confides in his son as they survey their field: "Our souls are buried down there," and extending the dandelion allegory to promote a perverse relational land ethic, Tadgh is told by his father to never trust a woman who has no contact with the land. Referring directly to the travelers, he perpetuates a common Irish myth that they lost their footing on the land during the famine and how they will never get back on it.

This opening credit sequence quickly develops the central story of the film when Tadgh, playing the gormless if unintentionally cruel son, asks his father why the widow (Francis Tomelty) does not want to sell the field. His father responds how she is a woman, like your mother, "sure if we knew what would keep them happy, we'd be in paradise." Reminiscent of a historical strand of insecure paternalism,[16] while displaying an inherent envy toward the female's more natural, nurturing, and life-creating instincts, the farmer needs to sublimate such love and creativity and transfer his emotional empathy to his field through nurturing and hard work, drawing seaweed to enrich his beloved soil.

When the Bull confronts the publican about the sale of his precious land, which he has rented for so long from the widow, he is warned that "outsiders" might bid for the prize possession. This activates a conventional reaffirmation of postcolonial and national solidarity, evoking victimhood and the rise of a common enemy: "who drove us to the coffin ships?" Raising the communal passion of the pub crowd, he rhetorically continues, "I drove them out . . . [the English]. No outsider will bid for the field."

Then, like all classical narratives, the plot thickens, and an outsider arrives to this insular farming community in the form of a returned Yank (Tom Berenger), presenting an inversion of the more benevolent trajectory of *The Quiet Man*. At an American wake, consisting of a celebration party for a native who is about to leave for America and most likely never return—the consequence of such mass emigration along with the famine, was to scar the historical memory and the psyche of the nation—the local tinker girl (Jenny Conroy), with conventional, flaming red hair taunts the men in the hall to dance with her. Correctly reading the situation but without the subtlety of a Jane Austen parlor mating game, the Bull fixes up Tadgh with his preferred partner, who has a suitable dowry and dances with great verve with the tinker girl. But poor Tadgh, who lacks many social skills, overdoes the dancing and frightens his intended partner and then quickly flees from the scene in disgrace.

At the auction the inevitable happens, and the Bull does not get his beloved field for the reasonable exchange value of £50. Instead, the widow doubles the asking price. Exploring the motivation of the Yank for buying the land, we observe him in an ostentatious automobile, surveying and witnessing the landscape with the local parish priest (Sean McGinley), both already on first-name terms. The Yank speaks glowingly of all of the limestone in the hills above and of his plans to cover the field with concrete to allow easy access to the area.

This perverse, materialistic desire for landownership, only to deny its primary life-giving use, is further dramatized by the tragic and fatalistic belief of the Bull, that "without land the Irish are both barren and dispersed," which corresponds to his own family trauma and enforced emigration. This central theme is affirmed by the promotional taglines on the DVD version: "[I]t owes him. It possesses him. It could even destroy him." Defined primarily as an economic resource, the land effectively broke up families while providing a meager subsistence for those left at home. Because of the scourge of economic emigration and the reality that there was only a living on the land for one family unit, this is suggested as the reason their elder son, Seamie, committed suicide almost two decades earlier.

The Yank's plan for the extreme exploitation of land is reaffirmed at a scene beside the local river when, rather than accepting less invasive pastoral harmony or land ethics, he raises the prospect of also harnessing the river for electricity with its potential for jobs. Meanwhile, the priest observes the mountains and recognizes beauty, as is normative in conventional romantic evocations of nature. Unlike the simple personal desire of Thornton in *The Quiet Man* to escape from his trauma, following an unfortunate boxing fight in America and return home to his roots, this Yank recognizes deep sadness and a lack of fulfilment in the natives. As an emigrant, he recognizes that the landscape is etched with centuries of barren wishes and dreams of how the land is unable to satisfy basic utilitarian needs. The priest concludes the discussion by asserting how the Yank's aims can be described as "noble," but the people do not like "change." Rather than citing a more pertinent, inherent land ethic, he simply restates the abiding trauma of a rural society where over a million died of hunger during the famine period and millions of others were forced into exile in the decades that followed. Very much addressing Sheridan's ambiguous attitudes toward the Irish mythos, the priest prophetically affirms his fears that "there's just a thin veneer of Christianity painted on these people."

In a later scene, where the Bull confronts the priest and the Yank, the Bull affirms that pouring concrete on fertile green grass is a mortal sin. The priest contests this apparently normative ethical assertion by revealing how the poor widow was terrorized for ten years, inferring how she therefore deserves retribution. While shocked at these revelations of actions secretly carried out by his son, the Bull nonetheless continues, moving from a strictly utilitarian and moralistic typology to an extremely humanitarian one by personalizing the land and speaking of it as his child that he nursed. The boundary between productive stewardship and a pathological variety is further illustrated by the narration of a primal scene from his childhood, which uncovers his buried and guilty consciousness. While working the fields trying to bring in the harvest with fears of summer rain that could destroy the crop, he witnessed his mother keel over and made the sign of the cross while she was dying. His father asked him to fetch a priest. Displaying his pathological credentials, he suggested that they should bring in the hay first, before attending to this sacred ritual. With this extreme response and with tears of pride, his equally pathological father realizes that the boy would protect their birthright at all costs. If he has "to face his mother in heaven or hell," then he cannot now let the field go becomes the punch line. Raising the temper of his speech even more for the priest and the Yank, the Bull rages that no (religious) collar, uniform, or weapon will protect the man who steals his land. This is probably as far from a holistic, benevolent manifestation of a land ethic as one can get. Ruth Barton concludes her review that while this "great Lear-like, Bull McCabe is the mouthpiece for romantic nationalism, he is also cast as tragic hero, brought down by his own grand delusions."[17] Barton, however, downplays the more primitive, even unnatural, love of the land, which I myself witnessed growing up on a family farm in the Irish midlands in the 1960s.

The night before the auction brings the crisis to a conclusion beside a beautiful, romantic waterfall. The Bull is there with his Lear-like staff framed against the glistening water, where he asks his son Tadgh to defend the family honor and teach the outsider the "lesson of the land." But Tadgh is not up to it and gets badly beaten by the Yank, reiterating his perceived lack of physicality and overall lack of ability as earlier registered by his parents. Like a ferocious father figure displeased with his offspring, the Bull grabs both men and throws them into the water after knocking their heads together. Continuing to bang the Yank's head across a rock, while cursing his interference, the Bull exclaims how he will "not be shamed," appearing not to recognize his awesome power. Later, crying out to God for forgiveness for what he has done, he all but howls at the moon. Looking down at this unnatural sight, one calls to mind the opening twilight scene that shielded the disposal of a dead donkey.

Even when the news of the Yank's murder becomes public knowledge, the local community refuses to betray the evil deed. In response, the local Catholic priest, apparently aligned with the forces of "perverse materialism" and the subversion of their authentic (pagan) identity, declares that such hunger for land will destroy their souls. "Through your silence," he pontificates from the church pulpit, you are supporting a murderer. He pleads with his flock not to add hypocrisy to cowardice. In an unconventional fit of anger, the priest locks the gates of the church until justice is done, while screaming at everyone to get out. Observing the proceedings, the Bull brazenly speaks of the natural injustice of it all and how "no priest died during the famine, only poor people like us."

Nonetheless, the Bull can no longer handle his psychological trauma, which is explained by his wife (Brenda Fricker), who at last speaks up after years of silence. She diagnoses that he has remained traumatized by guilt over their dead son for the last eighteen years. In a mythic, overtheatrical dénouement, he descends into uncontrollable madness and drives his beloved cattle over the edge of the cliff. Meanwhile, his son, trying to save the situation, also dies in the ensuing stampede. Following these tragic consequences, the Bull, embodying the tragic role of "King Lear," cradles his son's body on the rocks below, surrounded by his dead cattle and livelihood. "Kanute-like, he walks into the sea symbolically attempting to stop the tide of nature, no longer able to control his destiny or his sons'."[18]

The Field remains a primal text for exploring Irish national identity and the political and historical "land question" that has plagued the island for centuries. While this reading of the film tries to foreground a more universal ecological discourse around a farmer's ethical and legal relationship with his land, *The Field* also works at a mythical level and ensures that the personification of the Bull and his love of the land will live on in popular consciousness. Indeed, the story has continuing resonance across geographical space and time.

Concluding Remarks

This eco-critical analysis of these cautionary tales helps illustrate that while mythic farmers in *The Secret of Roan Inish* learn to unconditionally accept their holistic connection to land and nature, in more dramatic and problematic narratives such as *How Harry Became a Tree* or *The Field*, erstwhile nurturing if troubled protagonists are unable to do so and become extremely traumatized and psychotic as a consequence. These farmers, fixated on legalistic aspects of ownership and control alongside other more pronounced psychological and social problems not addressed here, were unable to appreciate that an ecological ethic must, according to Val Plumwood, always be an "ethic of eco-justice that recognizes the interconnection between social domination and the domination of nature."[19] Connecting such an abstract eco-critical ideal to the specifics of an Irish filmic aesthetic that is more directly associated with and dominated by a postcolonialist discourse cannot be fully unpacked through these short readings, but this chapter has attempted to do so nonetheless.

John Barry asserts that the majority of people in modern society "have no direct transformative experience of nature,"[20] having little direct connection to the land, except as some dramatic natural disaster. However, the evocation of ecological debates as personified through these films can hopefully sow the seeds of an alternative discourse that stresses good stewardship and green citizenship and at least promotes an awareness of our interdependence with our environment. I agree with Donna Haraway, who says, "We must find another relationship to nature besides reification, possession, appropriation, and nostalgia."[21] Unfortunately, as we have seen, Irish representations of farming and landscape are frequently codified and reduced to these anti-ecological attributes. A new nature aesthetic and an ecological preoccupation need to be constantly postulated to keep ecological debates high on the cultural agenda of film study and analysis.

Notes

1. This type of topography is contrasted to manicured colonial gardens, designed primarily for easy engagement with nature, as exemplified by Edmund Burke's study "Beauty and the Sublime." Luke Gibbons's superb study of Burke's treatise written in Ireland explores the dichotomy between the beautifully manicured and wild nature outside of the enclosed aristocratic estate. "The complexity of Burke's theories of violence, sympathy and pain, as outlined in his "A Philosophical Enquiry in the Origins of our Ideas of the Sublime and Beautiful" (1757), [when he was just twenty-seven—but begun at least ten years before when a student at Trinity] provided him with a set of diagnostic tools to probe the dark side of the Enlightenment, particularly as it was used to justify colonial expansion, religious bigotry, or political repression" (Gibbons, *Edmund Burke and Ireland* [Cambridge, MA: Cambridge University Press, 2003] xi).

2. There are few, if any, full-blooded examples of dystopic narratives around farming in Ireland, however. The only one that comes to mind is the recent generic horror film *Isolation* (2005), directed by Billy O'Brien, which stars John Lynch as an almost bankrupt farmer who sells his cows for genetic experimentation, all of which of course goes horribly wrong.

3. John Button, *Dictionary of Green Terms* (London: Routledge), 190.

4. Aldo Leopold, *A Sand County Almanac and Sketches Here and There* (New York: Oxford University Press, 1949), 224–25.

5. Stephen R. Kellert, in a paper in Richard L. Knight and Suzanne Riedel, eds., *Aldo Leopold and the Ecological Conscience* (New York: Oxford University Press, 2002), 133.

6. This highly contested typology of values associated with Wilson's biophilia hypothesis, as suggested by Kellert in an essay in the edited volume *Aldo Leopold and the Ecological Conscience*, 134, can be recategorized to correspond to the Irish films examined here. Note, however, as James Joyce said about *Ulysses*, the structure is merely a loose scaffold and can be pulled away after its job has been done.

1. Aesthetic—physical attraction and appeal of nature. Film intrinsically develops this in varying utopian ways.

2. Dominionistic—mastery and control of nature. Westerns and other genres address this while it also is critiqued in revisionist narratives.

3. Humanistic—emotional bonding with nature. Displayed, for example, within the Disney aesthetic alongside documentary nature and romantic fiction films and underpinned by a poetic and romantic sensibility.

4. Naturalistic—exploration and discovery of nature. This can be seen in documentary and nature films.

5. Moralistic—moral and spiritual relation to nature. This is explored through a range of deep ecological films.

6. Negativistic—fear of and aversion to nature. This is often an urban/contemporary fixation, which at extremes considers the rural landscape backward and is evidenced in some (Irish) academic criticism of film.

7. Scientific—knowledge and understanding of nature. This is most usually foregrounded in documentary film.

8. Symbolic—nature as a source of communication and imagination. Again a romantic preoccupation one could argue.

9. Utilitarian—nature as a source of material and physical rewards. This is a dominant criticism raised against global capitalism and a pressure that can be recognized within much representation of nature and farming. In this study its abuse appears to influence the psychological well-being of the farmers concerned.

Kellart incidentally wrote how this hypothesized dimension of the "Biophelia tendency" may constitute the most compelling argument for a powerful conservative ethic. See http://www.prescott.edu/academics/adp/programs/scd/sustainable-ways/vol2-no, accessed March 17, 2007.

7. Friedrick Ratzel (1844–1904) often is considered the father of environmental determinism and first argued how the natural environment was immutable, underlying all human activity. In turn, Ratzel was influenced by George Perkins Marsh, who in 1864 published his very influential *Man and Nature* (just five years after Darwin's *The Origin of Species*), which considered humans active agents effecting change in the environment.

"Nature" within this discourse provides a rich reservoir of socially constructed, historical, and philosophical attitudes and beliefs concerning ecology. Most notably while the seventeenth-century philosopher, Thomas Hobbes, viewed the "natural human condition," prior to the emergence of civilized society, as "nasty, brutish, and short," his contemporary, John Locke, considered that nature embodied a state of "humanitarian bliss" and affirmed that "natural laws" must form the basis of a just society (cited in Anderson 1996, 5). Descartes, on the other hand, construed nature as "the Other" and in so doing reified nature as a thing—an external other—entirely separate from the world of thought. Nature became, as Heidegger later complained, "one vast gasoline station for human exploitation." Adorno, in *Revolt of Nature*, continued in this vein and prophetically contended that mastery over nature inevitably turns to mastery over men.

Cited in J. Anderson, *The Reality of Illusion: An Ecological Approach to Cognitive Film Theory* (Carbondale: Southern Illinois University Press, 1996) 134.

8. Leopold, *A Sand County Almanac*, 200–201.

9. Kerstin Ketterman, in the essay "Cinematic Images of Irish Male Brutality and the Semiotics of Landscape in *The Field* and *Hear my Song*," in *Contemporary Irish Cinema: From the Quiet Man to Dancing at Lughnasa*, ed. James MacKillop, 154. (Syracuse, NY: Syracuse University Press, 1999.

10. While often emulating conventional tropes of romantic landscape—most famously politicized for a nationalist agenda by W. B.Yeats—Kavanagh most successfully embodies the more concrete expression of the life of a farmer in classic poems such as *The Great Hunger*.

> Clay is the word and clay is the flesh
> where the potatoes are gathered like mechanized scarecrows.

Or, in his poem *Epic*, which often is used to validate the "local" as equally important in parallel with to "universal," contrasted with the more inward looking and disparaged "parochial." The poem also has obvious parallels to the local/universal narrative recounted in *The Field*.

> I have lived in important places, times
> when great events were decided, who owned
> that half a rood of rock, a no-man's land
> surrounded by our pitch forks . . .
> a local row. God's make their own importance.

He depicts the tyranny of landownership and his great poem *The Great Hunger*, according to Patrick J. Duffy, in the essay "Writing Ireland," which "reflects on the social and emotional wasteland that imprisoned in his farm has brought him, as well as the irony of the iconic urban myth of the peasant." Such pessimistic readings of farming are contrasted with "From Tarry Flynn," which represents a celebration of going to help at a threshing. See *In Search of Ireland: A Cultural Geography*, ed. Brian Graham, 70. (London: Routledge, 1997).

11. Cited in Kevin Rockett, et al., *Cinema and Ireland* (London: Routledge, 1988), 194.

12. As affirmed by Ruth Barton in *Irish National Cinema* (London: Routledge, 2004), 73. Its worldwide success coincided with the establishment of "The Irish Tourist Board" [Bord Failte] in July 1952 and "acted as a blueprint for subsequent travelogue films promoting Ireland as a tourist attraction. . . . In 1985 when it appeared on video it sold 200,000 copies in the first four years in Britain alone." Luke Gibbons, *The Quiet Man* (Cork: Cork University Press, 2002), 3–4.

13. Many cultural critics such as Declan Kiberd suggest that postcolonial Irish identity was redefined, if not measured, by opposition to the colonialist ruler of the country.

14. Leopold, *A Sand County Almanac*, x.

15. Collection of seaweed for use as fertilizer is recorded since early times in Ireland, and access to good seaweed was regarded as very valuable. Some varieties were (and are) used in the human diet and for cattle feeding. See Fergus Kelly's *Early Irish Farming* (Dublin Institute for Advanced Studies, Dundalk County Louth: Dundalgan Press, 2000), 406.

16. See Declan Kiberd, *Inventing Ireland*, who affirms that patriarchal values exist in societies, where men lacking true authority settle for mere power (New York, NY: Vintage Press, 1996), 390–91.

17. Ruth Barton, *Irish National Cinema* (London: Routledge, 2004), 135.

18. As cited by Ruth Barton, *Jim Sheridan: Framing the Nation* (Dublin: Liffey Press, 2002), 47.

19. Val Plumwood, *Feminism and the Mastery of Nature* (London: Routledge, 1993), 20.

20. John Barry, "Green Politics and Ecological Stewardship," in *Questioning Ireland: Debates in Political Philosophy and Public Policy*, ed. Joe Dunne et al., 146 (Dublin: Institute of Public Administration, 2000).

21. In the 1995 essay "Otherwordly Conversations, Terrains, Topics, Local Terms," cited in Pat Brereton, *Hollywood Utopia: Ecology in Contemporary American Cinema* (Bristol: Intellect Press, 2005), 55.

10

Postcards from the Andes

Politics of Representation in a Reimagined Perú

Teresa E. P. Delfín

> The subaltern, in other words, is not only *acted on*; . . . it *also acts*
> to produce social effects that are visible.
>
> —John Beverley, et al., *Latin American Subaltern*
> *Studies Group Founding Statement*

Introduction

While most would be hard pressed to name a single Peruvian visual artist, even those who have never visited South America can tell you what a picture of Peru looks like. Catherine Stimpson describes "the superficial representations of Andean life" as "a picture of a llama or of a smiling Indian child playing the flute" (Weismantel 2001, 19). But Andean life also looks like countless other things that have been excluded from the canon of Andean imagery. Certainly there are mountains, children, flutes, and llamas in the Andean landscape, but there also are rivers and deserts, shopping malls and skyscrapers, uniformed schoolchildren and denim-clad teenagers—all of which are at odds with the Andean visual archetype. While the dominant images often are dismissed as quaint or powerless—if not a reflection of the way things actually are—the archetype they inform has been instrumental in shaping a notion of the Andes that is desired and paid for by tourists and, consequently, banked on by the state. As such, the images do indeed behave rhetorically: they engage their viewers and argue, above all, for their authenticity, silently insisting that the world they represent is real and intact. There also is a powerfully charged metonymy at play in these images, with the pictures imposing an ideal to which they challenge the landscape to conform.

Where these visuals are most dominant is in tourism, where they are put to work to recruit consumers. Of course, as with any commodity, the site being sold is expected to be accurately reflected in the advertisements that attract consumers in the first place. In Peru, what was sold in recent years was a place out of time, where descendents of the Inca roam colonial streets peopled with nondescript locals and young European backpackers on holiday. And for several years there was (and might still be) a long-haired man with indigenous features who guarded the twelve-angled stone (*hatun rumi*) dressed as the Inca. Cusco's vast and not always connected tourism promoters became experts at promoting this ideal, leading to a surprisingly unified suggestion that what was being represented was really something as uncomplicated as "reality." Images of this version of Peru are everywhere—from travel brochures to clothing catalogs and from postcards to fine art photography.

These depoliticized and ahistorical fantasies conflicted with and challenged the dominant late-twentieth-century notion of Peru as a state perpetually threatened by a repressive government and radical insurgent groups. What they offered up instead was no longer a political state but a "land" outside of a geopolitical purview. Imaginatively, then, Peru's borders expanded and contracted as if in Alice's Wonderland: here Peru transgressed its own borders, becoming instead the more mythical *Andes*, there Peru was reduced to a pile of rocks: Machu Picchu. It is precisely these sorts of decontextualized imaginings—often promoted by the state in hopes of provoking historical amnesia in its beholders—that inform the canon with which I am concerned. These pictures ultimately work toward the recruitment of foreign tourists whose presence helped ensure the proliferation of more and more sites in Peru that are consistent with this ideal. Central to these representations are two motifs: Andean nature and *campesinos*. These two categories ultimately work to convey the same message and often are seen as one and the same. But where did they come from? What informed this specific genre, of all the possible versions of Andean iconography?

Peru is a mostly rural country whose governing bodies, historically, have had little in common with the citizens they have represented. The same is true for Lima, the colonial capital city, a bustling giant in comparison to Peru's other urban areas. Nonetheless, Lima had, since the colonial era, been thoroughly invested in putting its whitest face forward to the outside world. This was a vision of itself, a representation of the state, that was believed to be appealing in its colonial mimicry to taste-making Europe. But this has shifted notably in the last century. With the rise in photography came a gradual diminution in state representation by educated, wealthy *Limeños*. The ascent of the Andean image world (Poole) as the dominant regional rhetoric initially helped democratize perceptions of the state, contributing images of Indians, *mestizos*, and the countryside to an already bulging archive of portraits of the urban well-to-do. But the balance has tipped once again. Where nineteenth-century Peru was represented to the outside world as a disproportionately literary and Europeanized society, photography—the medium of choice of the last century—suggested that twentieth-century Peru was a doggedly rural, traditional society.

This chapter explores the history of the iconic *Cusqueño* imagery that first contributed to global fantasies of the Peruvian highlands and, nearly a century later, provided a proven model for the promotion of a perception of a benign state as an antidote to a time of violence and perceived danger. It is important that it was not painting, sculpture, architecture, nor, for that matter, the region's famous weavings or the astounding and enduring Inca architecture that became Cusco's most instructive and characteristic medium for visual representation; it was photography. And it also is significant that of the many important schools and styles of photography that were competing for limited clientele in the early twentieth century, it was the *indigenista* style of a single Cusco-based photographer that would continue to inform the normative Cusco image world for nearly a century. In what follows, I trace the rise of the iconic Cusco photograph, beginning with the *cartes de visite* that became popular in Peru in the late nineteenth century, not long after they were unveiled in Paris. I go on to trace the links between these *cartes* and the *indigenista* style they helped inform—especially in the first three decades of the twentieth century. From there I consider the impact this visual canon has had on *Cusqueños*, tourists, and the peasant subjects of the tourist gaze. As I argue throughout this work, the dedicated promotion in the 1990s and early twenty-first century of Cusco's canonical image world made long strides toward reestablishing a reputation of innocence and rustic tranquility in the aftermath of Peru's recent war.

The Limits of the Literary

Photography shifted the authority of universal knowledge from print language to spectacle.

—Anne McClintock, *Imperial Leather*

Throughout much of the twentieth century, there existed two dominant, but conflicting, visions of Peru to which the world outside of
the Andes was exposed: one imparted by text and the other by images.
News headlines, political documents, U.S. State Department warnings,
ethnographies, and novels involving Peru during this long period overwhelmingly depicted a poor, politically divided, and terrorism-tattered
state. In contrast, the Peru that was disseminated visually was one in
which nature was central, and people were presented as the timeless
keepers of ancient secrets.[1] Textual Peru came across as elitist, cryptic,
or violent. Throughout most of the twentieth century, the written word
in Peru was not intended for its masses or the outside world. Within
national borders, Spanish—especially in its written form—was the language
of colonial elites, not of the overwhelmingly illiterate Quechua- and
Aymara-speaking majority. And for the written word to circulate globally
in any substantial way would require translation. In contrast, the images
of Peru that circulated the globe at the same time presented what many
have interpreted to be a silent and suggestible landscape (Poole 1997;
Starn 1999; Weismantel 2001). In such images, indigenous people and
the natural environment are presented as interchangeable, possessing
parallel virtues. In a discussion on Andean postcards, Mary Weismantel
gets right to the point, explaining, "In the Andes, almost all the pictures
are of people or mountains" (2001, 179).

Historically, access to power and agency in Peru has been closely
linked to Spanish literacy and textual production. And although there
have been significant developments in regional subaltern styles, these were
mostly ignored by publishing houses and universities, which privileged
European forms.[2] In contrast, visual production—when it has been recognized at all—has been assumed to be outside of the realm of rhetoric;
that is, it has not been seen as capable of persuasion. Weismantel's recent
project is a case in point. She has begun to catalogue, photograph, and
interpret a remarkable collection of *Moche* ceramics from Peru. That
the pieces in these collections, which include elaborate, ironic, humor-
ous, and even erotic motifs, have languished in museum storage speaks
volumes about the neglect of indigenous artistic production in Andean
studies. To play off Gayatri Spivak's famous challenge, "Can the subaltern speak?" there is a long history of Andean high society responding,
"Of course not, but it can make pretty pictures." The trajectories of
agency and literacy in Latin America are so deeply interconnected that
in its founding statement the Latin American Subaltern Studies Group
noted that nationalism "is an elitist venture conducted by the same elite
guided in part by a 'literary' idea of nationhood" (Beverley, Aronna,
and Oviedo 1995, 144).

While Peruvian textual discourse has largely emerged from Lima or, on Peru's behalf, from Europe or the United States, many of the images that have come to be associated with Peru—and even, more generally, the Andes—have come from Cusco. Isolating visual culture from the competing textual hegemony, I analyze the strategies that have been deployed by *Cusqueños*, anthropologists, tourists, and even governmental agencies as a means to portray the rural sensibilities historically excluded by cultural producers in Lima. The types of images I discuss serve various purposes, but they all in some way promote the version of Peru that was hard-won in the nineties and became officially sanctioned by the Toledo administration. These images challenge the dominant late twentieth-century notion of Peru as a state perpetually threatened by a repressive government and the Maoist insurgent group, *el Sendero Luminoso*.[3] What they offer up instead is a timeless, depoliticized Peru as the land of Inca culture and abundant nature. The images I discuss are those that first presented Peru to curious Europeans and Americans in the nineteenth century and that later provided the visual road map to follow in the process of state reinvention. Just as *cartes de visite* in the *fin-de-siècle* catalogued "types" appealing to foreigners intrigued by exoticism, souvenir photographs canonizing Andean motifs would take hold a few decades later. These oversimplified and idyllic images came to be so thoroughly circulated worldwide that they provided an obvious and immediately persuasive visual argument for a Peru outside of political turmoil and terrorists' strikes.

When I began my fieldwork, I was surprised at how little the souvenir market in Cusco had changed from chronicles going back to the first half of the twentieth century. Most countries with well-established tourist economies have long since perfected the art of the souvenir, such that the typical trinket is read globally as a metonym for the destination itself. Leis represent time spent in the Hawaiian islands, as undeniably as oversized sombreros equate with trips to Mexico. But when I began my fieldwork in 2001, Cusco's newly revived tourism industry had yet to develop such a symbol—a fact that was duly noted by one of my informants—a European tourist who had invited me to go souvenir hunting with him as his trip was drawing to a close. He hoped I, as an anthropologist and thus a de facto "regional expert," would be able to recommend a must-have souvenir from a well-hidden shop. (This was before the large handicrafts market was established and before virtually every storefront in Cusco had something to offer a visiting tourist.)

The tourist was determined to find a souvenir that would "capture" his experience of Peru: a high-altitude wonderland of varied ecosystems, smiling peasants, rosy-cheeked children (which he described as "the world's cutest"), stunning archaeological ruins, colorful markets that, at the time, still mostly catered to locals, llamas, and once-in-a-lifetime trailside bonding with new and old friends. After spending the better part of a day ducking in and out of shops trying on alpaca hats and sweaters, comparing one rag doll to the next, and even holding up to the light commercially available slides (no doubt to pass off as one's own during slide shows back home), another tourist suggested that his own photographs, and not some odd trinket, would best capture the experience. He had kept his pockets filled with coins throughout the trip and never missed a chance to photograph any of the *campesinas* in *pollera* skirts or children in *chullu* hats that dutifully posed for a tip. He decided to process his film in Cusco and, following the clerk's recommendation, had the prints sepia-toned to give them the antiqued, rustic effect that is characteristic of Cusco's marketing efforts.

It is no exaggeration to say that in Peru, literacy—perhaps even more than gender or ethnicity—has determined access to political power. Until 1979, the right to vote was granted based not on the more common categories of landownership or gender but on the grounds of Spanish literacy. In a country where the majority of the population spoke Quechua or Aymara as a first language, the constitution's literacy clause all but ensured that power would remain in the hands of the country's Europeanized elite. Even after the inclusion of a cultural-sensitivity amendment banning literacy prejudice, it is widely recognized that reforms did not properly take effect until the 1990s. As a result, the overt literacy bias in Peru contributed to the subjugation of the majority of the country's citizens while assuring that texts remained conceptually fixed as a modern (read: colonial), elite medium, meanwhile leaving the country's poor literally speechless. Liisa Malkki's work on refugees—another category of extensively visually documented "speechless" people—provides a powerful model for making sense of this disjuncture. She writes, "Photographs and other visual representations of refugees are far more common than is the reproduction in print of what particular refugees have said" (Malkki 1996, 386).

Some have blamed the long-standing frustration with the Europhilic ruling class in Lima with the election of Alberto Fujimori, the no-name candidate who ran against literary superstar Mario Vargas Llosa. While Vargas Llosa's campaign was grounded in clear agendas for his administration, his opponent ran—and won—on a platform that leveled criticism against Peru's elitism by promoting himself as an average Peruvian. Fujimori's Japanese ancestry became both confused and conflated; his campaign nickname, "el chino," rendered him generically Asian, which in Peru meant, more importantly, non-white. And his consistent critiques of the elite ruling class made obvious his own outsider status in these circles—a status he shared with most of Peru. The battle for the presidency of Peru waged between Vargas Llosa and Fujimori brought to the fore the connection between literacy, race, and political power in Peru. Facing the possibility of being ruled by a president unimaginably different from themselves—the handsome, educated author at times seemed to have stronger ties to Europe than to Peru—*cholos* and *campesinos* rallied behind the unlikely hero, *el chino* Fujimori, seen by many as the lesser of two symbolic evils, appealing in his very averageness.

While this textual hegemony only increased throughout the last century, there were limits. Literary production in Peru had always been geared toward the privileged classes, and its very elitism limited its impact. In contrast, visual media had long experienced much greater mobility across divisions of class, language, and even nationality. As Deborah Poole shows in her excellent study of images in the Andes, what she calls the "image world" is far more inclusive and legible than its textual equivalent. *Campesinos* with little use for books, explains Poole, treasure photographs. These are wrapped carefully and stored in niches, framed and propped, and pasted onto walls, as are magazine and calendar cutouts, which serve similar purposes (1997, 4). These homegrown visual types, Poole's "image worlds," emphasize "the simultaneous material and social nature of both vision and representation." She goes on to note that "the specific ways in which we see (and represent) the world determine how we act upon that world and, in so doing, create what that world is" (1997, 7).

It is heartening, then, that recent *Cusqueño* literature has begun to come to terms with what we can consider both a rift and a richness between and among the regions' literary and image worlds. It is noteworthy that even the tireless defender of Cusco's letters, Luis Nieto Degregori, describes the characters that people Cusco's recent literary works in visual terms and according to their role in the *Cusqueño* visual pantheon. It is as though regional literature has been so momentously trumped by images that in order to coexist it has to reckon with the visual world, which owes much of its existence to tourism. Nieto, fittingly, describes Cusco's literary characters as "picturesque": "the backpacker and the hippy, the artisan who travels from city to city and from country to country, the amateur photographer loaded with cameras, [. . .] those who sell postcards and crafts on the streets" (Revilla 2001).

The Rise of Andean Photography

Before there were postcards or snapshots, the circulation of *cartes de visite* helped initiate a photography craze in Peru, much like it did in Europe and elsewhere in Latin America. These *cartes*, developed by Frenchman André Adolphe Eugène Diderí and patented in 1854, made mass production of photographs significantly easier and more cost-effective than it had ever been. And their diminutive size of 6 × 10 centimeters lent them a quality at once ephemeral and collectible (Gersheim 1988). If the popularity of these photos was any indication, then Susan Stewart was indeed onto something in linking the miniature, the souvenir, and the collection. Predictably, in a country of dramatic inequality, unabashed racism, and a perennial eagerness to appeal to Europe, these *cartes* had multiple faces and purposes in Peru's colonial and postcolonial eras. Among Lima's elite, the *cartes* served as calling cards, *tarjetas de visita*, on which messages and signatures were scrawled. Many collected the cards in albums—some especially designed to house these collections—as evidence of their broad social networks. In the Andes, the *cartes* craze was such that in 1886 an estimated 1 million of the palm-size cards were produced in Colombia alone, a country of fewer than 4 million inhabitants (Levine 1989, 26). *Cartes* served to establish class status and profession, to advertise and announce, and to catalogue and amuse. They were commissioned by everyone, from politicians to prostitutes.

Eventually, *cartes de visite* would cease to be produced in sole service of promoting or presenting their depicted subjects. The small portraits would eventually focus increasingly on " 'typical' or exotic figures from society—water carriers, 'savage Indians,' old men and women, beggars" (Levine 1989, 28). Following the ethnological model, the subaltern subjects of these photos were lit, dressed, and staged in ways that conferred upon them artificial timelessness, authenticity, and dignity. Unlike the *cartes*, individuals commissioned for their own enjoyment, which were mostly portraits or seated poses of a single person or at most two people, the *cartes* marketed to tourists and focused on local types tended to show anonymous people—sometimes several on a single card, filling the tiny frames. Photographers' lenses were backed considerably away from their rural models, as though seeking to distance themselves from any real interaction with their subjects while ensuring the best vantage. Anne McClintock astutely reframes early photographic practice within the framework of surveillance, making the camera analogous to the panopticon. "The camera," she explains, "embodies the panoptic power of collection, display, and discipline" (McClintock 1995, 123).

Unlike the true calling cards, these tourist versions of the *cartes* no longer focused on faces. The subjects of these cards were never meant to be recognized as individuals, only as types. In all cases, the *cartes* were staged photos of passive, patient subjects, captured in indoor studios. The portraits of wealthy clients depict a consistent look of boredom, the subjects propped in mock salons implying bourgeois backgrounds. In contrast, the cards made for commercial distribution highlighted the exotic, "backward" lives of locals with whom visiting Europeans were so taken. The purpose of these cards was as souvenirs proving contact with the exotic characters they depict. The models in these photographs are frequently burdened with props from their workaday lives, staging meant to naturalize the subjects, as though captured spontaneously. A young man's hand inside his *ch'uska*—a bag meant for carrying coca leaves—suggests a momentary break from hard work in the field. A woman posing in profile with a bundle on her back gives the impression that she and her presumed baby are on their way somewhere.

The captions beneath the carefully catalogued *cartes* collected by Dr. L. C. Thibon, the Bolivian consul in Brussels, highlight the differences between the types of *cartes* that circulated in the Andes. Individuals' cards are annotated with names and sometimes anecdotes or genealogies: "Mariquita, President Pacheco's mistress," "The beautiful Rosine," and "Doña Juana Vidaure, arch-millionaire who sometimes wears more than 2,000,000 jewels" (Poole 1997, 125). The "Indian types" collected by Thibon, while likewise captioned, never name the individuals. Their annotations serve only to establish the "Indians' " role in the labor force, or to illustrate what Thibon perceives to be cultural traits. The characters in these *cartes* are variously described as "man who sells corn beer," "agriculturalist," and even the audacious "*La Toilette*: The women eat the lice which they find" (Poole 1997, 128). McClintock is again instructive, suggesting that "as a technology of surveillance, photography was central to the rationalizing of working-class leisure and work time. As such, it was associated with those other panoptic Victorian phenomena—the exhibition, the museum, the zoo, the gallery, the circus—all of which involve the fetishistic principle of collection and display and the figure of panoramic time as a commodity spectacle" (1995, 123).

Inspired by the images that had begun to circulate throughout Europe and the United States, many early tourists to the Andes in general and Cusco specifically were drawn to the promise of interacting with—or at least seeing evidence of—cultures radically different from their own. And as remains the case with today's tourists, those early travelers acquired souvenirs that were at times more about proving than memorializing their visits. Certainly *cartes de visite* filled that niche, but their ephemeral nature left a desire among many travelers for something more substantial. Imported tastes created a market in Cusco for high-quality photographs of regional motifs. Unlike the staged ethnological images that first began circulating in Europe, bent on displaying the subjects' "savagery" and inferiority to the white photographers and collectors, many of the images that came out of Cusco in the first decades of the twentieth century were deeply indebted to the *indigenismo* movement that got its start at the University of Cusco. The project of *indigenismo*—simultaneously social, political, and aesthetic—was invested in valuing indigenous contributions at all levels of society. This included efforts toward reforms in workers' rights, renewed appreciation of the region's early indigenous cultures, and the development of artistic creation that simultaneously rejected European influence and established regional styles. In reaction to colonial pressure to *mestizar*—to reject indigenous heritage and adopt European and colonial manners—some *indigenistas* took to wearing traditional clothes—*pollera* skirts, wool hats, and vests—and the use of Quechua saw an urban resurgence that continues to this day (Deustua and Rénique 1984).

The influence of the *indigenistas* provided a counterpoint to the Europhilic taste that dominated Lima at the time. While the capital was celebrating modernism and the avant garde alongside Europe, Cusco's *indigenista* community bucked the trend toward elitism. Instead, Cusco's cultural commentators (of which there were few) championed homegrown styles celebrated for having emerged organically, in contrast to the studied styles popular in Europe and Latin America's urban capitals. The practitioners of local styles of European art forms—analogues to European and American bohemians—were described by some as *walaychu*, wanderers, who replace "family and community and tradition with a deeply sentimental attachment to the land [which is] credited as the source of the *waylachu's* heightened artistic and musical sensibilities" (Poole 1997, 177–78). The Quechua term *waylachu* inherently implied provenance in the high Andes, and rightfully so. A figure like the *waylachu* could never have come from or existed in Lima; the insistence on differentiating Cusco and Lima ran multidirectionally, with each side policing its own boundaries against influence from its perceived dark side. The *waylachu*, however, like the *brichero*, was not a single type to which artists tried to conform. Instead, the archetype provided a general description along a spectrum that contained both a connection to nature and a proclivity for the arts. In *indigenismo's* highly reactionary political climate, the *waylachu* paradigm allowed artists to ignore the rules of society, family, and community while still being recognized and included in all of these social structures. The artist interpreted by his community as a *waylachu* could continue along his way without resorting to foreign models for validation.

Two of the Cusco photographers active during the peak of *indigenismo*, Juan Manuel Figueroa Aznar and Martín Chambi, could each have been described as *waylachus*, And though they apprenticed in the same studio, their work was dramatically different. The son of a successful father employed in the mining industry and a Spanish mother, Juan Manuel Figueroa Aznar was born in a small town in the mostly rural department of Ancash but was raised in Lima. It was there he received his fine arts training. And although the *mestizo*, middle-class Figueroa Aznar had managed one of the best arts educations available in Peru, his better resourced and more connected peers were studying in Europe and returning to Lima to acclaim. Though he lacked European anointing, Figueroa Aznar nonetheless tried his luck as a portrait painter and occasional photographer in the highly guarded elite circles in Lima. When he finally abandoned the capital, it was to work his way through Colombia, Ecuador, and Panama as a photographer. Once he returned to Peru, this time to the southern highlands, he took a job at the photographic studio of Maximiliano T. Vargas, who would later become best known as mentor to Martín Chambi. In 1904, at age twenty-six, Figueroa Aznar moved to Cusco, where his formal training granted him quick access to the town's small privileged class, and his lack of European affectation appealed to the growing *indigenista* circles.

A little more than a decade later, Martin Chambi, like his predecessor Aznar, would make his way from the Max Vargas studio in Arequipa to Cusco. The son of indigenous Peruvian *campesinos*, Chambi lacked the privilege and education that marked his soon-to-be colleague, but he benefited from a fortuitous encounter with a photographer when working in a mine as a boy. The experience was so powerful that Chambi dedicated himself to the study of photography and quickly showed sufficient talent to be accepted as a young apprentice at the Max Vargas studio. In the early 1920s, Chambi moved to Cusco, where he would make a career of photographing the notable events in the lives of the regions' elites, and—ultimately, more significantly—taking his camera out of the studio to document the everyday lives of ordinary people and the goings-on of the former Inca capital.

On the surface, the similarities between the two photographers, Figueroa Aznar and Chambi, are considerable: Both photographers trained under Vargas. Though neither was a native *Cusqueño*, both made their careers and permanent homes in Cusco. And, most importantly, both made an impact on the field of photography through their treatment of Indians as agentive subjects. Both were among the few photographers working in Cusco at the time, a time that coincided with the rapidly growing popularity of photography worldwide, as well as the "discovery" of Machu Picchu. As a result, both photographers were, to some extent, charged with representing Cusco. Inspired by the *indigenista* undercurrent that by the time of Chambi's arrival in the 1920s would have been hard to miss, both created images that were self-conscious social statements. The work of the two was seen as compatible enough to have been included in the same exhibit and accompanying publication in 1934, *Cuzco Historico*. Both photographers made names for themselves as Andean artists dedicated to advancing regional realities through their work.

But whereas Chambi was a self-made man who worked for a living—usually by producing conventional studio portraits of the few in Cusco who could pay for the luxury—Figueroa Aznar came from a good family and married into an even better one. Aznar's comfortable life afforded him the luxury to create art for art's sake, which in his case resulted in far greater experimentation across media and styles. Given his background in fine arts, growing interest in theater, and a theoretical bent that is apparent in many of his works, his oeuvre was less accessible than Chambi's. But as much as Aznar pushed the envelope of Peruvian photography, most critics have noted that Chambi's technical skills were unquestionably superior to his colleague's. Chambi's photography was simply world-class.

While Chambi is only now achieving name recognition—often mentioned among the ranks of such photographers as Tina Modotti and Walker Evans—his pictures have for decades enjoyed great circulation, mostly as postcards. The ubiquity of Chambi's postcard images is such that they are still in print today—not as self-consciously historical images but simply as photographic examples of Peru. Further, the phenomenon of the Chambi postcard has even been immortalized in text, making an appearance in the work of *Cusqueño* author Luis Nieto. In his short story, "Buscando un Inca" ("In Search of an Inca"), the main character, a *Madrileña*[4] named Laura Cristóbal[5] is traveling around Cusco, Peru, in search of an authentic Andean experience. Fed up with the pushy anthropologists and promiscuous shamans that a *turista* must brave in search of Peruvian authenticity, she reconsiders the goals of her trip. Nieto writes:

> Cured of her anthropological inclinations—of her zeal to meet history face-to-face—she decided to return to the fold and came to know Cusco like just any other tourist, brandishing her camera while a swarm of trinket-sellers followed here everywhere. She also bought several sets of postcards by the Peruvian photographer Chambi and managed to convey in a few lines the illusion that she was getting along just fine. The encounter that finally turned things around for her occurred as she was scribbling

away on one of these—the one with the family playing the toad game—in the Café Varayoc (2002).

Sapo, that "toad game," is still ubiquitous in Cusco. No doubt as a result of the classic Chambi photo, visits to taverns indistinguishable from that in the picture have become popular tourist destinations, and a game of *sapo*—which consists of attempting to throw a coin into the mouth of a brass toad—has likewise persisted as a quintessentially *Cusqueño* pastime. Organized city tours in Cusco, without overtly referencing the Chambi postcard, nonetheless reproduce the motif daily, stopping briefly at *chicherías* for tourists to try their hand at *sapo*. It is interesting that although I have played the game myself while doing research about these tours, in all the time I spent in *chicherías*, I never played—or saw anyone playing—outside of the context of a carefully scripted tourist activity. There is irony in this. A defining image by the definitive photographer of indigenous Peru has now become part of Peru's pandering to tourists' perceptions of authenticity. Here, as in other aspects of the representation of both individual and state identity in Peru, it is not indigeneousness itself that is privileged but a recreation of scenes from a tourist imaginary. But whatever degree of performance may be said to be found in these tourist tableaus does not detract from the place of Chambi in the pantheon. On the contrary.

Argentinian photographer Sara Facio wrote of Chambi, "We are talking of the first Indian photographer. Martín Chambi is a descendant of the Incas. The first great photographer who sees his people through uncolonized eyes" (in Chambi, et al., 1988, 11). Indeed, we must wonder if it is Chambi's own standpoint—*un indio campesino* like his subjects—that makes his photos appear so iconic, so natural, so unguarded. Chambi's connection to his subjects is so complete, in fact, that without the help of captioning it is nearly impossible to distinguish his self-portraits from his pictures of other *indios*. Figure 10.1 and 10.2 both depict a type he made famous—the *indio* in the mountains. The only reliable clue we get as to which one is the self-portrait is that the photographer is wearing boots, whereas the *nativo* is barefoot, as his *campesino* subjects frequently are.

The eloquence and sincerity of Chambi's images have such an impact that it not only transforms those who view them, it also has permanently impacted the landscapes and people that were his muses. The effects of Chambi's images have been far reaching, instructing tourists on what to desire, suggesting to Peruvians how they should appear, and creating a template from which few photographers in the Andes have strayed. As Mario Vargas Llosa explains, "The world Chambi

Figure 10.1. Niña y llama T. Delfin

Figure 10.2. Tourist Gaze T. Delfin 2001

photographed so indefatigably, [. . .] he also transformed. He placed his own personal seal upon it, a grave order, a ceremonious and somewhat iconic air, an immobility that has touches of the disturbing and the eternal" (Chambi, et al. 1988, 7). Although Chambi retired more than a half century ago, his photos continue to teach us how to see, how to

be, and what to want from Peru, for better or worse. As Weismantel
has noted, the timelessness that is part of the much-mimicked Chambi
model racializes the subjects of the photographs (2001, 180). As has
become common, tourism has picked up where anthropology had the
sense to leave off—in a world of its own design, where out-of-the-way
places are synonymous with racialized "pastness."

While Peruvian textual production has overwhelmingly excluded
Peru's marginalized peasant majority—both as consumers and as protagonists—*campesinos* and the natural landscape with which they often
are linked have become the darlings of Andean visual rhetoric. But
unlike literary products, images of Peru are predominantly produced by
and for consumers outside of Latin America—namely, tourists—making
their appeal far broader, and arguably more compelling. Through global
distribution of postcards, JPEGs, Web sites, and advertisement images,
visual Peru has quickly eclipsed its textual counterpart in its impact on
the global imaginary. Peru's visual turn has been central to the reinvention of the state as a refuge—on a global scale—from the chaos of late
capitalism. The silently communicative visual language of advertisement
images, postcards, guidebooks, snapshots, Web sites, and coffee-table
books presents the Andean state as the antithesis to the noise, chaos,
pollution, congestion, commodification, media saturation, speed, secu-
larism, and—counterintuitively—the fear of violence that has become
rampant in urban centers in the twenty-first century. In domestic terms,
the imaginative possibilities presented by these images have worked to
make a case for Peru as safe and peaceful. As a result, Peru has almost
completely shed its image as the land of the Shining Path and Tupac
Amaro Revolutionary Movement (Movimiento Revolucionario Tupace
Amaro), replacing this reputation with the appealing and globally recognized images of Machu Picchu and happy peasants.

People/Nature

Two "types" dominate the visual culture of contemporary Peru, the
postcard-perfect shots of women and children in traditional, vibrant
wool dress and the relentlessly recurring vistas of green-and-granite
Machu Picchu. With men conspicuously absent from most images, the
two dominant visual categories often are packaged to appear as organically intertwined, repeating visually the tired connection between nature
and the feminine. But whereas Machu Picchu often stands alone in
images, women are commonly captured so as to appear as part of the
landscape. Indeed, at the center of the Peruvian visual genre is an effort

to reproduce images that can be read, above all, as natural—candid, unstaged, everyday images. Just as the images are frozen moments in time, the argument put forth by "classic" images of Peru is of a land and people that are likewise static, unaffected by colonialism, terrorism, neoliberalism, or globalization.

The reality, of course, is that each paradigmatic image tells a story of the social conditions that have informed it. It is no accident that the protagonists of Andean visual production are overwhelmingly peasants. And of this already marginalized group, it is indeed the most subjugate who are perpetually in the camera's frame: women, children, and the elderly. Because these are the people whose access to employment and education is most limited, their own provincialism becomes a prized and marketable attribute. To this end, expensive tours promise to provide "an inside look at 'real' rural Peru" (STA 2001, 31). The women and children who have become the faces of Peru are idealized by foreign media—and consequently foreign tourists—as being untainted by the stuff of modernity. But not only does one's presence or absence in these photogenic landscapes reflect contemporary social conditions—it also has helped shape the conditions themselves.

It is no surprise that men have all but disappeared from recent photography in the style Chambi helped canonize. Scholars of Latin America have noted that men often are the first to lose their ties to traditional culture—often because they must leave their rural homes to make a living. In the areas surrounding Cusco, work is often sought in the tourism industry. Most often, employees in the highly competitive tourism industry represent themselves as multilingual and well educated, often possessing advanced degrees in tourism or related fields. There are, of course, exceptions in which men (and women) seek urban employment that depends on their self-representation as indigenous. *Bricherismo* is perhaps the best example of this and also an ironic inversion of both the normative economic and rhetorical models. *Bricheros*, and also porters, are ubiquitous characters on the Cusco tourism circuit who most often come from surrounding rural areas or promote themselves as such. When they leave, their wives, sisters, children, and little brothers remain to make whatever living they can. When these villages attract any foreign attention, locals are accustomed to collecting a few coins in exchange for having their picture taken. These photos are carefully shot by tourists familiar with what the genre consists of—namely, dusty *campesinos* with no trace of modernity surrounding them. One at a time, the images help reinforce the myth that Cusco, Peru, and the Andes are "just like that."

Just as tourists have always been motivated by expectations of what their destinations have to offer, so have the objects of their gazes learned to anticipate tourist desires, and to use them to their advantage. Knowing that they constitute part of a marketable landscape, *campesinas* and their children have been leaving their rural homes—either during the workday or for good—to try their luck with the transient population that values them most: tourists. It has now become common enough to see *cholas*[6] walking past high-heeled businesswomen that social scientists and citizens alike now speak of the *indigenización* of the metropolitan capital, Lima. Where Lima used to be an enclave for modern elites and the modernizing poor, it has increasingly come to reflect the state's diversity. Contemporary Lima can be read as an indication of the success of Peru's latest reinvention as a postmodern indigenous state. The attempt to realize this new identity is such that the slogan for the government's national promotion Web site is "Perú: Land of the Inkas."[7] Not only is the indigenous population being made central in this rhetoric, it is even selling its own *indigenización* as complete, by using the Quechua spelling of Inca, *Inka*.

Where *campesinas* have been most effective at connecting themselves to the desires of tourists is by simply making themselves present in tourism sites. Whether or not they are genuinely *of* the spaces to which they link themselves, wherever a tourist bus might stop, one can count on seeing traditionally clad women and children (and sometimes their likewise photogenic baby sheep or goats) waiting to have their pictures taken. The purpose of their placement in these landscapes is, above all, to profit on the desirability of their image. Archaeological ruins crowded with foreign tourists are not authentic in and of themselves. But an indigenous presence in the same might suggest to visitors that these spaces are still valued and used in historically continuous ways, thus giving visitors the impression that they are not just taking in a site but also participating in it. Of course, the tourism experience is not complete until there is documented proof of the experience, and for this tourists are willing to pay.

Figure 10.1, an example of the sort of "tourist gaze" one can witness at any major tourist destination in and around Cusco, depicts a young girl with her llama. Tourists who are unaware that she is wearing the highland equivalent of her "Sunday best" might assume that she is simply en route between one place and another as part of her normal day's business. This photo, sent home in an e-mail, or, if featured later in a slide show depicting a Peruvian vacation, might be presented as proof of an excursion into "real" Peru. Widening the lens, however, shows that the girl is literally surrounded by gawking, photo-snapping tourists. What is so easily cropped to tell one story can just as easily be seen as part of a broader, and quite possibly terrifying, context. As part of her day's work, this girl and the many like her, must turn herself into a spectacle and submit to being encircled by throngs of strangers who hover at least a foot above her. Her work is also illegal, and she knows that she can be arrested and even beaten if she is discovered pestering tourists without a license to sell goods. And there is no question that a tourist's work, in southern Peru, is always worth more than that of a *campesina*, whom we can assume, does not speak Spanish, let alone English or German, the dominant languages of Peruvian tourism. Nonetheless, at the end of the impromptu photo shoot, she will collect money in exchange for her image. And while she is young enough to attract a tourist's gaze, she is considered fortunate to be able to make a little money in a less dangerous or degrading way than other members of her community.

As I have attempted to illustrate in this chapter, representations of Peru often are cast simply as nature-as-it-is, with people indigenous to said "nature" regarded as flora or fauna, and rarely much more. Yet the realities they represent are far from inherent and are no more signs of what Peru may "really" be than are the equally available images of Bohemian backpackers, flea markets, or multinational chain restaurants. But it is worth lingering on the motivations—personal and political—for transnationally promoting a vision of Peru that is far more in sync with what is seen on postcard racks than what one experiences on the street. I maintain that the primary motivation is this: in response to terror, the most hopeful thing to do is to recast reality one fantasy at a time. As discussed earlier, because postcards teach tourists how to see, and thus locals how to be, there is no question that once a solid enough foundation is built, fantasies *do* inform reality.

Notes

1. In the conclusion to her famous *testimonio*, Rigoberta Menchú powerfully redeploys the silence that is expected of indigenous Latin Americans, writing, "I'm still keeping secret what I think no-one should know. Not anthropologists or intellectuals, no matter how many books they have, can find out all our secrets" Quosedin Bloom 147 (1998).

2. Inca Garcilaso de la Vega, one of Peru's most celebrated writers, wrote his entire oeuvre in Spain, never to have returned to the Peru that served as the staging ground for all of his works.

3. *El Sendero Luminoso*, the Shining Path, was founded in Peru in 1970 as a violent, class-based revolution that was especially active in the 1980s and 1990s. A 2003 commission estimated that during those two decades of violence, at least 30,000 people were either killed or disappeared by the insurgency. The

insurgency was quelled by the end of the 1990s, in part as a result of a major anti-insurgency campaign under then-president Fujimori (Eisner).

4. A woman from Madrid.

5. It should not be overlooked that the main character is Christopher Columbus's namesake, who in Spanish is called Cristóbal Colón.

6. "*Cholas* is the name the nonindigenous have adopted for those Indians who attempt to become mestizos and who migrate to the city" (Burt 2004, 21).

7. Commission for the Promotion of Peru, 2007.

Works Cited

Bloom, Leslie Rebecca. 1998. *Under the Sign of Hope: Feminist Methodology and Narrative Interpretation.* Albany, NY: State University of New York Press.

Beverley, John, Michael Aronna, and José Oviedo. 1995. *The Postmodernism Debate in Latin America.* Durham, NC: Duke University Press.

Chambi, Martín, et al. 1988. *Martín Chambi : Fotografías Del Perú, 1920–1950: Biblioteca Luis-Angel Arango, Octubre–Noviembre De 1988.* Bogotá, Colombia: Banco de la República.

———. 1993. *Martín Chambi: Photographs, 1920–1950.* Washington, DC: Smithsonian Institution Press.

Deustua, José, and José Luis Rénique. 1984. *Intelectuales, Indigenismo Y Descentralismo En El Perú, 1897–1931.* Cusco, Perú: Centro de Estudios Rurales Andinos "Bartolomé de las Casas."

Gersheim, Helmut. 1988. *The Rise of Photography, 1850–1880: The Age of Collodion.* London: Thames and Hudson.

Levine, Robert M. 1989. *Images of History: Nineteenth and Early Twentieth Century Latin American Photographs as Documents.* Durham, NC: Duke University Press.

Malkki, Liisa H. 1996. "Speechless Emissaries: Refugees, Humanitarianism, and Dehistoricization." *Cultural Anthropology* 11.3: 377–404.

McClintock, Anne. 1995. *Imperial Leather: Race, Gender, and Sexuality in the Colonial Contest.* New York: Routledge.

Nieto, Luis. 2002. "In Search of an Inca." Gowanus. (Winter 2002). http://www.gowanusbooks.com/brichero_eng.htm, accessed 13 June 2007.

Poole, Deborah. 1997. *Vision, Race, and Modernity: A Visual Economy of the Andean Image World.* Princeton Studies in Culture/Power/History. Princeton, NJ: Princeton University Press.

Revilla, Vicente. (Autumn–Winter). "Bricherismo: Romancing the Gringa: An Interview with Lucho Nieto." Gowanus. http://www.gowanusbooks.com/nieto-interview.htm, accessed 19 May 2006.

STA. 2001. "South America: Sta Travel Style." Ed. GAP Adventures.

Starn, Orin. 1999. *Nightwatch: The Politics of Protest in the Andes.* Durham, NC: Duke University Press.

Stewart, Susan. 1993. *On Longing: Narratives of the Miniature, the Gigantic, the Souvenir, the Collection.* 1st paperback ed. Durham, NC: Duke University Press.

Weismantel, Mary J. 2001. *Cholas and Pishtacos: Stories of Race and Sex in the Andes.* Women in Culture and Society. Chicago, IL: University of Chicago Press.

11

That's Not a Reef.
Now *That's* a Reef

A Century of (Re)Placing the Great Barrier Reef

Kathryn Ferguson

Imagine a world that comes into being when nature becomes tech-
nologized beyond even these wild dreams.

—Rosaleen Love, *Reefscape: Reflections on the Great Barrier Reef*

The Great Barrier Reef has been an oversized silent partner in Australian
history—it has played a significant role in elections, military conflicts,
social movements, reconciliation and repatriation, the arts, and some of
the world's most advanced and groundbreaking science. The reef is the
largest natural feature on the face of the planet and is the basis of the
second-largest marine protected area in the world. It is an important World
Heritage site—covering 347,800 square kilometers and comprising some
2,900 individual reefs and 900 islands. It is home to six of the world's
seven species of turtles; 16 species of sea snakes; over 200 species of
birds; 400 species of coral; 500 species of seaweed; 1,500 species of fish;
and 4,000 species of mollusks. Approximately thirty species of dolphins,
porpoises, and whales travel through or reside in the reef's waters. The
reef is the largest system of coral reefs and associated life-forms in the
world and is one of the world's most diverse ecosystems.

The Great Barrier Reef is one of those iconic places that seemingly "everyone" knows something about. They may know that it can
be seen from space, that it is a World Heritage site, that it is vulnerable
to global warming and pollution, that there are some 2,000 shipwrecks

in its waters, or simply that it is somewhere around Australia, where Nemo lives. This sundry range of information and knowledge has, for the most part, not been gleaned from serious concentrated study but from representations of the reef in popular media.[1] This dissemination has been effective. "When people overseas are asked to name the one thing that they associate with Australia, 83% name the Great Barrier Reef" (Hoegh-Guldberg and Hoegh-Guldberg 2004, 23). The domestic attachment to the reef was confirmed in 2003, when an Australian Electoral Commission survey of Australians found that 93.6 percent of respondents wanted greater protection for the Great Barrier Reef (GBRMPA 2003). The worldwide reputation the reef casually commands is understandable, as the reef has featured in countless documentaries, popular science journals, photographic essays, tourist touts, movies, news stories, postcards, posters, works of fiction, and even computer games. Obviously, the Great Barrier Reef is not everyone's idea of "the environment." Nor should it be, although in thinking of "the environment," far too often we overlook the vast submerged ecosystems that make up a dominant share of our world. This may be due in part to the fact that we tend to translate "the environment" to "*our* environment," which, for the most part, is limited to land. However, the Great Barrier Reef is one unique oceanic space that in an increasingly urbanized and technologically driven new economy has been invested with value. The reef is a unique biological and geological entity, and, to use Sharon Zukin's (1991) phrase, a cultural "landscape of power," where different visions of nature, different images of science, and different arguments of humanity's proper place in the natural world, each with its attached cultural and political implications, are played out.

In his book *The Age of Extremes,* Eric Hobsbawm (1994) highlights the difficulty of writing contemporary histories: "If the historian can make sense of this century," he contends, "it is in large part because of watching and listening" (3). In making sense of Australia's Great Barrier Reef, however, it is particularly what we *see* of the reef—sleek cruising sharks, majestic staghorn coral, rhythmic garden eels, technicolor cuttlefish, elusive prehistoric nautilus, and oddly adorable Maori wrasse and potato cod in crystalline waters—that makes us *believe* that Australia's Great Barrier Reef is indeed great. However, only a relative few individuals will ever witness firsthand what happens, or has happened, on the reef.[2] For most of the world, "making sense" of Australia's Great Barrier Reef is largely dependent upon an uneven and idiosyncratic collection of images that has been mediated by various graphic technologies and presented in an assortment of pictorial forms. We "know" the reef, and we "experience" it through pictures of subjects from which we are spatiotemporally distant.

As Susan Sontag (1984) points out in *On Photography*, "Reality has always been interpreted through the reports given by images" (153), but in the middle of the nineteenth century, with the invention of photography, "the allegiance to images" was intensified to the extent that photographic images became a supplement rather than a complement to the real. Walter Benjamin (1978) notes that photography also brought this "allegiance to images" to a much wider audience, as it "enormously expanded the scope of the commodity trade by putting on the market in unlimited quantities figures, landscapes, [and] events that [had] either not been salable at all or [had] been available only as pictures for single customers" (151). Sontag goes on to comment that, phenomenologically, we are increasingly interpreting reality through the images presented to us by popular media, much as we interpret images from our own experiential reality. Or, as Barthes (1981) expresses the same sentiment, with photographs the "past could be as certain as the present, what we see on paper is as certain as what we touch" (88). As Gerhard Richter argues, "The photograph is the only picture that can truly convey information, even if it is technically faulty and the object can barely be identified. A painting of a murder is of no interest whatever; but a photograph of a murder fascinates everyone (in Obrist 1995, 56–57). The photograph is believed to have an ontological connection to reality that allows the information contained within its surface to be interpreted as part of our knowledge of the "real world."[3] We can "see" and "know" something—experience some persistent quality of the moment—from a photograph disconnected from that moment by both time and distance. As Sontag contends, photographs not only "give people an imaginary possession of a past that is unreal, they also help people to take possession of space in which they are insecure" (1984, 9). Historically, this has persistently been the case of the Great Barrier Reef, as its remoteness from Eurasia and the Americas and the fact that it is, for the most part, underwater have meant that although the reef is well known, it is primarily known from a distance and mediated through various photographic technologies.[4] Thus for most of us, the Great Barrier Reef is effectively a virtual reef; an aqueous collage of images garnered from a diversity of photographic sources.

W. J. Thomas Mitchell's (1994) assertion, that we need to think more seriously about images, about what they are, "what their relation to language is, how they operate on observers and on the world, how their history is to be understood, and what is to be done with or about them," (13) offers a critical opening—an intellectual wedge—to begin addressing visual representations of the Great Barrier Reef. This chapter takes up Mitchell's invitation to raise a critical question in terms of the Great Barrier Reef: If the world knows the reef predominantly through visual images, and we already have a vast collection of those images, then do we still need the reef? This is not an idle musing. According to *Australian Geographic*, the coral cover of the reef has declined by half over the last forty years (Woodford 2004, 47), the Melbourne *Age* front-page headline on February 12, 2005, succinctly announced in large typeface that it was already "Too Late to Save the Reef," the *Sydney Morning Herald* joined the list of mourners eulogizing "The Late Barrier Reef," and among the more dispirited of marine biologists, the reef is being muttered about as the "So-So Barrier Reef." Professor Ove Hoegh-Guldberg, director of the University of Queensland's Center for Marine Studies, who leads an international study on reefs and climate change for the World Bank and UNESCO, released a report in February 2004 that suggested that the reef, in a best-case scenario, will not survive a lot longer due to the impact of global warming, rather than agricultural and urban runoff, tourism and development pressures, or overfishing, which also are placing increasing pressure on the reef. Due to the effects of increased ocean temperatures, within twenty years, what may well be presented to people who travel to see for themselves the wonders of the reef will effectively be Great Barrier Reef "theme parks," wherein "a reef" will be isolated and cultivated, and interactive computer screens will simulate the expected "reef experience" (Dickie and Brown 2004). Sadly and ironically, the Great Barrier Reef is in the process of being entirely replaced by its own image. However, what we are seeing visually represented as the Great Barrier Reef is too often not the reef at all; it is a molded and marketed commodity that has eclipsed the reef itself.

This use of photographs to represent—supplement and replace—the reef, which may be distant in both time and space, is not an entirely recent development. In 1891, the Commissioner of Fisheries for Queensland and former president of the Royal Society of Queensland, William Saville-Kent, sent a series of black-and-white photographs to Burlington House in Piccadilly. There the photographs were exhibited in Saville-Kent's absence by Sir William Flower, Fellow of the Royal Society and director of the Natural History Museum. For the first time, without abandoning the dubious comforts of Victorian London, both professional and amateur naturalists could see for themselves some of the corals and other living wonders of the Great Barrier Reef. Two years later, Saville-Kent (1893) would publish sixty-four of his photographs in *The Great Barrier Reef,* and images of the Antipodean wonder were quite suddenly available around the world.[5] As James Bowen and Margarita Bowen (2002) note, Saville-Kent's inclusion of meticulously detailed "chromo" images of sixteen reef inhabitants within his opus "marked a revolutionary approach to introducing the general public to the incredible beauty and variety of reef life" (159). However, as Rosaleen Love (2001) observes, Saville-Kent's "reefscapes," which depict the reef at uncommon interludes of low tide and glassine calmness on the windward side of the reef, "framed beauty in such a way that it proved—not deceptive exactly—but scientifically misleading."[6] Indeed, as Love goes on to note, Charles Maurice Yonge, leader of the British Association for the Advancement of Science's Great Barrier Reef Expedition, in 1928–1929, found Saville-Kent's photographic reefscapes "far from typical," although the members of the expedition had left England with their expectations of the reef derived almost exclusively from Savile-Kent's thirty-five-year-old photographs (Yonge 24, quoted in Love 2001, 101).

Over a century after Saville-Kent's first hand-colored images of the Great Barrier Reef were presented to the world, we are still facing the same conundrum confronted by Yonge and his expeditionary team: How extensively have photographs of the Great Barrier Reef created unrealistic expectations of it? For example, in a 2003 episode of Radio National's "The Science Show," in a panel discussing the survival of the Great Barrier Reef, Richard Fitzpatrick remarked that

> shooting for the BBC Australasia series about the Great Barrier Reef . . . we cannot find a lot of locations suitable to meet the aesthetic quality that the broadcasters want, so in fact a lot of the filming we've been doing over the last couple of years we actually have to perform out in the Coral Sea rather than in the GBR lagoon because of the quality [of the Reef] itself. (qtd. in Williams 2002)

Fitzpatrick, a Townsville-based marine biologist and cinematographer, freely acknowledges that much of the footage, which is presented as being representative of the Great Barrier Reef in his work and the work of his peers and may actually be images of the reef's kilometers outside of the Great Barrier Reef region, is troubling—especially as it is due to the fact that parts of the Great Barrier Reef no longer *look* that "Great."[7] This fact becomes a particularly vexing irony when articles sounding jeremiads for the destruction of the reef, such as the aforementioned *Australian Geographic*, *Sydney Morning Herald,* and *Age* articles, are incongruously accompanied, if not dominated, by beautiful color photographs of a vibrant, ostensibly healthy, reef.

Postcards and Hunting Trophies

The dream touches the region where pure resemblance reigns. Everything there is similar; each figure is another one, is similar to another and yet to another, and this last to still another. One seeks the original model, wanting to be referred to a point of departure, an initial revelation, but there is none. The dream is the likeness that refers eternally to likeness (Blanchot 1982, 268).

Because going either beyond the Reef or back in time are the only ways that parts of the Great Barrier Reef can still be visually sold as "Great," stock footage of animals long dead, coral decimated years ago, and crystalline water from decades past is being recycled—well beyond their subjects' use-by dates. It is nearly impossible to know if Fitzpatrick's sharks have been filmed in the Great Barrier Reef region, or if they reside some 150 kilometers east of the Great Barrier Reef Marine Park at Osprey Reef. It is equally difficult for the casual observer—the purveyor of postcards, the online virtual user, or the package trip traveler, for instance—to know that pictures taken in 1982 of a reef site boasting healthy coral colonies and a darting multitude of vibrant fish may well now only offer a barren oceanscape—a stark and almost colorless reality of coral bleaching, algae overgrowth, or crown of thorns starfish. This is not to say that these aged photographs are not useful. Dated images of the reef are being used by biologists as aids in retrospectively calculating the population levels of various species. However, despite their very real value as graphic testimony to population fluctuations and topographical changes on the reef for specialists, photographs of the reef in and of themselves, although offering us the information that at some time, at some place, these specific situations were available for photographic imaging, do not necessarily convey to the casual view of the nonspecialist enough information to be indexically connected to a still-extant reality.

Chronologically ambiguous images of the reef, however, cannot be dismissed as being without value. Many of these images, however empirically problematic, are stunningly beautiful. They offer aesthetic pleasure to their viewers, among whom I must number myself. I am a committed virtual user and voyeur of the reef, both past and present. Indeed, if I recall correctly, it was a *National Geographic* article chock-full of pictures that convinced me, at about age five, that I needed to visit the reef and "save all the animals." It has not quite worked out that way. However, it was the incredible and entirely "other" beauty of the reef and life on it depicted in those big, glossy photographs that initiated my investment in its well-being—and might well be cited as the aged origin of this chapter. Love reports that my visual induction and continued fascination are not unusual. In her experience, when asking people "why they have come to the Reef—including Reef scientists—they always say 'It's beautiful'" (2001, 10). It is, at least in part, because the reef is beautiful that we have invested it with value and do not want to see it destroyed. As Judith Wright noted in her renowned history of the early efforts to save the reef from the catastrophic effects of limestone mining and oil drilling in the late 1960s and early 1970s, the story of those early efforts to save the reef and preserve it as a marine park "is the story [of] the battle to save that thousand-mile stretch of incomparable beauty" (1977, xiv). Indeed, in being listed as a World Heritage site in 1981, the reef is noted as containing "unique, rare and superlative natural phenomena, formations and features and areas of exceptional natural beauty" (2005). Since most of us have seen that exceptional beauty, or at least a part of it, through the work of photographers and filmmakers, it might follow that the work done by the visual images of the reef is a germinal to its preservation. If we did not know it was beautiful—if we could not see for ourselves that the reef is magnificent—then would we still want to preserve it? Would the accuracy of scientific arguments of biodiversity, ecosystems, and environmental conservation and preservation influence us as effectively as a somewhat ambiguous photograph or film footage of extraordinary beauty?

The reef that most of us know as beautiful has, through its history, been cited as both a rich resource and an awkward obstruction. As Saville-Kent pointed out in his presidential address to the Royal Society of Queensland in 1890, "The Great Barrier Reef is rich beyond imagination, in the production of a marine fauna redundant with forms possessing both an economic and scientific value" (qtd. in Bowen and Bowen 2002, 158). The wealth of the reef, which Saville-Kent did his best to protect from resource raiding, was, even in 1890, imperiled by overfishing, especially of bêche-de-mer and pearl-shell. On July 1, 2004, the zoning of the Great Barrier Reef Marine Park was modified, so rather than less than 5 percent of the reef being designated as a marine sanctuary, now roughly one third of it is designated a "no-take zone." As well as fishing and related industries, by the 1960s the coral itself was being considered a valuable source of dredged limestone, and the ocean floor around the reef was marked as a highly favorable potential site for oil drilling. Queensland's state mining engineer, a supporter of the drilling, suggested that Heron Island be used to test the "theory" that oil spills would do nothing to the coral reef environment. This experiment would see the dumping of numerous barrels of oil into the waters around the island to see what happened to the marine environment (Clare 1971, 11). It was not sensible state legislation that stopped oil drilling on the reef (not a surprise, as State Premier Joh Bjelke-Peterson held a major interest in Exoil-Transoil, one of the forty groups of oil companies holding drilling rights on the reef) but a "green ban" threatened by several trade unions that were expected to service offshore drilling rigs (Lawrence et al, 2002, 34–35). In 1975, oil drilling was finally prohibited on the reef. The suggestion that nuclear explosions be used to clear parts of the reef for the convenience of large ships carrying minerals for export was bandied about in federal parliament between 1967 and 1968 (Clare 12–13). Currently about 4,500 ships greater than fifty meters in length (mostly bulk carriers) still use the passage inside the reef, bringing with them the threat of potential groundings that bear the concomitant threat of the release of large, concentrated quantities of oil and chemicals, including tributyltin and copper sulfates, into the reef environment.

Among the very early conservation efforts to control and curtail destructive resource extraction and commercial exploitation on and around the reef, E. J. Banfield's work was arguably the most politically effective. Edmund James Banfield was a rather introspective journalist, fond of Thoreau's *Walden Pond*, who, under the pen name Rob Krusoe, wrote a series of articles on the reef for *The Bulletin*, a Townsville newspaper (Bowen and Bowen 2202, 219). In 1897, Banfield moved to Dunk Island, where he would live for twenty-six years, and wrote *Confessions of a Beachcomber* (1908), *My Tropic Isle* (1911), and *Tropic Days* (1918). As Lawrence, Kenchington, and Woodley point out, Banfield's texts "helped promote the natural beauty of the region, stimulate public imagination and popularise the romance of life on the tropical north coast," while most contemporary and subsequent publications extolled the commercial values and advanced economic exploitation of the reef (2002, 22). Banfield's *Confessions* has never been out of print—the pecuniary texts, which included a children's book, have faded into obscurity. Here I have wandered somewhat afield of my discussion of visual images of the reef, but the point I want to draw from the example of Banfield's authorship is that even from the very early beginnings of reef conservation efforts, it would be the political power of the perceived *beauty* of the reef that swayed, and continues to sway, popular opinion—in the face of, or perhaps despite of, other more prosaic uses of the reef and its waters. This is no small accomplishment.

Michael Pollan's *Botany of Desire* puts forward an intriguing theory that what humans perceive as the "beauty" of nature is an overlooked characteristic in the ongoing selection process for survival, and that the human desire for beauty forms a part of natural history. Contending that the genetically "successful" species, which had survived into the twenty-first century, "which [he had] always regarded as the objects of [his] desire, were also . . . subjects acting on [him], getting [him] to do things for them that they couldn't do for themselves" (2001, xv).

Pollan explains:

> All it requires are beings compelled as all plants and animals are, to make more of themselves by whatever means trial and error present. Sometimes an adaptive trait is so clever it appears purposeful: the ant that "cultivates" its own gardens of edible fungus, for instance, or the pitcher plant that "convinces" a fly it's a piece of rotting meat. But such traits are clever only in retrospect. Design in nature is but on concatenation of accidents, culled by natural selection until the result is so beautiful or effective as to seem a miracle of purpose. (2001, xxi)

Following Pollan's line of reasoning, then, it may well be arguable that the beauty of the reef—disseminated throughout the world by photographers and filmmakers to a world that would otherwise not know of its underwater wonders—has served to preserve it. But if we are, at least in some ways, desirous of preserving the beauty of the reef, and that beauty is mimetically available to us through visual images, then is preserving its visual beauty too easily imagined as preserving the reef itself?

Photographs always refer to something that has preceded them and are thus never the origin but supplement the origin, exceed the origin, and may well survive the origin. In *Camera Lucida*, Roland Barthes (1981) is enthralled by the ambiguous status of photographs as both an elusive fragment of time suspended in a temporal ever-alive present and a tangible artifact of the deadness of the past: "All those young photographers who are at work in the world, determined upon the capture of actuality, do not know that they are agents of Death. This is the way in which our time assumes Death: with the denying alibi of the distractedly "alive," of which the Photographer is in a sense the professional" (92). The connection between photography and death emphasized by Barthes is echoed in a slightly less macabre tone by Sontag: "One of the perennial successes of photography has been its strategy of turning living beings into things, things into living beings" (1984, 98). Peter Schwenger neatly sums up the argument: "Attesting to the reality of the object depicted, the photograph implies that it is alive, even if that life is always in the past tense and thus already dead" (2000, 396). Indeed, Benjamin attributes the success of the photographic medium, at least partially, to this characteristic ability to blur life and death: "The cult of remembrance of loved ones, absent or dead, offers a last refuge for the cult value of the picture. For the last time the aura emanates from the early photographs. . . . This is what constitutes their melancholy incomparable beauty" (1968, 226). Loss haunts the photograph. The photograph of a "real live" fish on the Great Barrier Reef may well be, to use Barthes's words, "the living image of a dead thing. For the photograph's immobility is somehow the result of a perverse confusion between two concepts: the Real and the Live: by attesting that the object has been real, the photograph surreptitiously induces belief that it is alive" (1981, 79). Have scenic vibrant photographs and films of life on the reef undermined our ability, our willingness, to accept the very real and very pressing reality that what constitutes the beauty of the reef is dying?

I am not proposing here a grand conspiracy theory in which photographs and films are being intentionally deployed by a shadowy cohort to deceive the masses, or that there are no longer any places on the reef where life abounds and is marvelously spectacular. What I am proposing is that the imaginative geographies created by the profusion of available images of the reef create an illusory photographic continuum that occludes temporal and spatial gaps. Thus, following Kracauer's 1927 argument of representations of natural disasters, a meaningless, if not false, historicism is constructed by the "flood of photos" that "sweep away the dams of memory" (1993, 432).[8] For example, André Bazin suggests that the pictographic arts perform the same function as embalming the dead, rendering every collection of photographs a triumph over decay (1967, 9). However, when visiting a natural history museum, the preserved specimens on display are, quite obviously, dead. As Martyn J. Linnie points out, "Many specimens in [natural history] museums today date back several hundred years" (2000, 296). We are offered a clear indication of what a specimen may have looked like during its life—but there is no doubt that the specimen might decide to leave the display and go about its former life. However, a Nemo-esque clown fish photographed in, say, 1991, is not likely to be an extant *Amphiprion ocellaris* today—but as a postcard image, it will *look* like it is, or at least could be. The death of that particular clown fish is eternally suspended—perpetually deferred, but not for the clown fish—only within our human understanding.

This does not mean that the image is somehow inadequate to presenting visual information, but, in Blanchot's words, "tends to withdraw the object from understanding by maintaining it in the immobility of a resemblance which has nothing to resemble" (1982, 260). The creatures and topography of the reef are not, in this way, unique—wildlife and landscapes no longer in existence are the subjects of countless amateur and professional photographic efforts. The underwater "otherness" of the reef, however, leaves it particularly vulnerable to visual fetishism. By this I mean that because most of us will never see the reef for ourselves, discrete images of it are taken as indicative, as visual supplements, of the whole—all 400,000 square kilometers of it. Much as incongruous collections of colonial "artifacts" were used to represent India, not as it was but as it was *imagined* to be, at Eurocentric world fairs in the eighteenth and nineteenth centuries (see Breckenridge 1989), photographic artifacts of the "other" and thus "exotic" reef tend, as saleable economic commodities, to reflect only the fragments of the reef that equate with what we *want* the reef to be like. We want the reef to be beautiful and vibrant; untrammelled by human incursions and influences. I have yet to see a postcard showing bleached coral or dead sharks.

In the process of writing this chapter, I have been debating the possible inclusion of photographs. I want to show you what the reef looks like—*convince* you, *prove* to you, that it is beautiful in a way that words cannot. Conversely, I also want to offer you stark images of bleached coral, anchor damage, and dead fish. But I am leaving the visual images to your imagination. Did you skim this chapter before reading its text to see if there were photographs of the reef? Did you look for the pictures first? Were you disappointed that I did not include any? Perhaps you imagined that photographs of the reef might reveal you a truth that I would not.

Notes

1. Indeed, a 1999 Australian survey discovered that television was the overwhelmingly dominant source for public information regarding the Great Barrier Reef. See Green, et al. 1999.

2. Between 1994 and 2004, some 17,642,372 individuals visited the Great Barrier Reef Marine Park. See Great Barrier Reef Marine Park Authority 2005, http://www.gbrmpa.gov.au/corp_site/key_issues/tourism/gbr_visitation/QuickBar/reef_query.cfm?section=reef&sectionTitle=Reef%20Wide&scale=.00025, accessed August 21, 2005.

3. The notion of photographic "evidence" has been used formally and informally to both prove and disprove a myriad of theories. Everything, from the existence of faeries in Cottingley to the success of NASA's Apollo moon missions, has been "proven" and "disproven" on the basis of photographs.

4. Sadly, this is too often the case for many of the recreational divers who *do* visit the reef. Being entirely invested in their roles as amateur photographs and taking (usually bad, boring, or blurry) pictures of the reef, their experience of the reef is limited entirely to their viewfinders—and reviewing the pictures topside.

5. In the usual rather long-winded Victorian fashion, the full title of the large quatro was *The Great Barrier Reef: Its Products and Potentialities, Containing—An Account with Copious Coloured Illustrations and Photographic Illustrations*

(the latter here produced for the first time) of the Corals and Coral Reefs, Pearl and Pearl-Shell, Beche-de-Mer, Other Fishing Industries, and the Marine Fauna of the Australian Great Barrier Region. A second edition of the volume was produced in 1900, and in 1972 a facsimile edition was published.

    6. Love 2001, 101. Love's title of "Reefscape" echoes Saville-Kent's own description of his photographs of Great Barrier Reef corals at low tide defining the horizon.

    7. Fitzpatrick's work has been seen on, among others, the Discovery Channel, the National Geographic Channel, ABC, BBC, CBC, and Japan's NHK and TBS. He has received numerous underwater cinematography awards but is perhaps best known in Australia as the "shark tracker" for his ongoing research on several species of sharks.

    8. I do not believe it is too far a rhetorical stretch to examine the current situation of the reef in terms of a natural disaster. Although not the abrupt destruction of a tsunami or a hurricane—the gradual death of the reef is no less catastrophic.

Works Cited

Australian Government Department of Environment and Heritage. 2005. "Great Barrier Reef World Heritage Values." http://www.deh.gov.au/heritage/worldheritage/sites/gbr/values.html, accessed October 13, 2005.

Barthes, Roland. 1981. *Camera Lucida: Reflections on Photography.* Trans. R. Howard. New York: Hill and Wang.

Bazin, André. 1967. *What Is Cinema?* Trans. Hugh Gray. Vol. 1. 2 vols. Berkeley: University of California Press.

Benjamin, Walter. 1968. "The Work of Art in the Age of Mechanical Reproduction." In *Illuminations,* ed. Hannah Arendt, 217–51. New York: Shocken.

———. 1978. "Paris, Capital of the Nineteenth Century." In *Reflections: Walter Benjamin: Essays Aphorisms, Autobiographical Writings,* ed. Peter Demetz, 146–62. Trans. Peter Demetz. New York: Schocken.

Blanchot, Maurice. 1982. *The Two Versions of the Imaginary.* Trans. Ann Smock. Lincoln: University of Nebraska Press.

Bowen, James, and Margarita Bowen. 2002. *The Great Barrier Reef: History, Science, Heritage.* Melbourne: Cambridge University Press.

Breckenridge, Carol A. 1989. "The Aesthetics and Politics of Colonial Collecting: India at World Fairs." *Comparative Studies in Society and History* 31:2: 195–216.

Clare, Patricia. 1971. *The Stuggle for the Great Barrier Reef.* London: Collins.

Dickie, Phil, and Susan Brown. 2004. "The Late Barrier Reef." *Sydney Morning Herald,* February 21.

Fyfe, Melissa. 2005. "Too Late to Save the Reef." *Age* 12 (February): 1, 10.

Great Barrier Reef Marine Park Authority. 2003. "Representative Areas Program: Protecting Our Great Barrier Reef." http://www.gbrmpa.gov.

au/corp_site/management/zoning/rap/rap/protecting_our_gbr.html, accessed August 22, 2005.

———. 2005. "Reef Wide Total Visitors by Year." http://www.gbrmpa.gov. au/corp_site/key_issues/tourism/gbr_visitation/QuickBar/reef_query. cfm?section=reef&sectionTitle=Reef%20Wide&scale=.00025, accessed August 21, 2005.

Green, D., G. Moscardo, T. Greenwood, P. Pearce, M. Arthur, A. Clark, and B. Woods. 1999. "Understanding Public Perceptions of the Great Barrier Reef and Its Management." Cairns, Australia. CRC Reef Research Center. Vol. Technical Report 29.

Hobsbawm, Eric. 1994. *The Age of Extremes: The Short Twentieth Century, 1914–1991*. London: Abacus.

Hoegh-Guldberg, Ove, and Hans Hoegh-Guldberg. 2004. "Time's Almost Up for the Great Barrier Reef." *Australasian Science* (April): 23–25.

Kracauer, Siegfried. 1993. "Photography." *Critical Inquiry* 19:3:421–36.

Lawrence, David, Richard Kenchington, and Simon Woodley. 2002. *The Great Barrier Reef: Finding the Right Balance*. Melbourne: University of Melbourne Press.

Linnie, Martyn J. 2000. "Prevention of Biodegeneration in Natural History Collections: Potential Trends and Future Developments." *Museum Management and Curatorship* 18:3:295–300.

Love, Rosaleen. 2001. *Reefscape: Reflections on the Great Barrier Reef*. Washington, DC: Joseph Henry Press.

Mitchell, W. J. Thomas. 1994. *Picture Theory: Essays on Verbal and Visual Representation*. Chicago, IL: University of Chicago Press.

Obrist, H. U., ed. 1995. *Gerhard Richter: The Daily Practice of Painting: Writings and Interviews 1962–1993*. Cambridge, MA: MIT Press.

Pollan, Michael. 2001. *The Botany of Desire: A Plant's-Eye View of the World*. New York: Random House.

Radio National (Australia). 2003. *Can the Great Barrier Reef Survive?* Radio National. Polly Rickard, January 11.

Saville-Kent, William. 1893. *The Great Barrier Reef: Its Products and Potentialities*. London: W. H. Allen.

Schwenger, Peter. 2000. "Corpsing the Image." *Critical Inquiry* 26:3: 395–413.

Sontag, Susan. 1984. *On Photography*. London: Penguin.

UNESCO. 2005. "World Heritage Center—The Criteria for Selection." http://whc.unesco.org/en/criteria, accessed June 24, 2008.

Williams, Robyn. 2002. "Tracking Sharks." *The Science Show*. Australia: ABC Radio National, November 2.

Woodford, James. 2004. "Great? Barrier Reef." *Australian Geographic* 76: 36–55.

Wright, Judith. 1977. *The Coral Battleground*. Melbourne: Thomas Nelson.

Yonge, Charles Maurice. 1931. *A Year on the Great Barrier Reef: The Story of Corals and the Greatest of Their Creations*. London: Putnam.

Zukin, Sharon. 1991. *Landscapes of Power: From Detroit to Disney World*. Berkeley: University of California Press.

Part 4

Seeing in Space and Time

12

Evading Capture

The Productive Resistance of Photography in Environmental Representation

Quinn R. Gorman

W. J. T. Mitchell's (1994) article "The Photographic Essay: Four Case Studies" begins with an air of discomfort: a nagging sense that within the academic discussion of photography as a form of representation something valuable is being passed over, dismissed in some ultimate and rather rash fashion, and that an overweening confidence is being placed in its dismisser. "The relationship of photography and language," he writes, "admits of two basic descriptions, fundamentally antithetical":

> The first stresses photography's difference from language, characterizing it as a "message without a code," a purely objective transcript of visual reality. The second turns photography into a language, or stresses its absorption by language in actual usage. This latter view is currently in favor with sophisticated commentators on photography. It is getting increasingly hard to find anyone who will defend the view (variously labeled "positivist," "naturalistic," or "superstitious and naive") that photographs have a special causal and structural relationship with the reality that they represent. (1994, 281–82)

One of those "sophisticated commentators" Mitchell cites is Victor Burgin, who asserts that all photography is "invaded by language in the very moment it is looked at," and who refers to the "*naive* idea of purely retinal vision" (qtd. in Mitchell 1994, 282, emphasis added). Further, Burgin contends that perspectives holding out some place for photographic exceptionalism are "relics [. . .] obstructing our view of photography" (ibid.). Mitchell is troubled by Burgin's assessment only in part, and hence his discomfort: while he accepts the notion that language can condition visual perception and cognition, and while he agrees that verbal and visual representation cannot be so easily isolated from one another, he has a very rhetorical concern over the certainty with which Burgin makes his claims (i.e., how that certainty may be a product of will rather than necessity).

In Mitchell's view, the rhetoric of an abject photography returns within Burgin's very language, reversing the invasion and making insistently visible the continuing conflict between the two views of the photography/language relation. That Burgin couches his argument within a rhetorical figure that he has already supposedly disproven (his "unobstructed view" mirroring the "unobstructed view" of photography itself) suggests, at the very least, according to Mitchell, "that the relics [of a nonlinguistic photography] are not quite so easily disposed of" (1994, 283). Even more troubling for Mitchell is that Burgin "seems not to consider that this invasion [of photography by language] might well provoke a resistance, some real motive for a defense of the nonlinguistic character of the photograph" (ibid.). The questions, then, that drive Mitchell's subsequent investigation are:

> How do we account for the stubbornness of the naive, superstitious view of photography? What could possibly motivate the persistence in erroneous beliefs about the radical difference between images and words and the special status of photography? (1994, 283)

To answer these questions with a tone less problematic than Burgin's, he poses two more questions in reply:

> What if it were the case that the "relics" which "obstruct" our view of photography also *constitute* that view? What if the only adequate formulation of the relation of photography and language was a paradox: photography both is and is not a language? (1994, 283–84)

I quote and summarize at length from Mitchell's article for two reasons. First, I feel that his analysis meshes quite well with the purposes of the present collection, in that he seems to be approaching the issue at hand in a thoroughly rhetorical fashion. By looking at the way in which discourse and power are articulated within the photography/language discussion, he delineates not only the rhetoric about photography (the rhetoric of what we might call photographic theory, or representation theory, as it addresses photography) but also the consequences that the rhetoric *about* photography has for the rhetoric *of* photography. Put another way, the ways that we discuss or understand photography enable and affect the ways in which photography itself can exist for us, with a concomitant influence on the intelligible message(s) of photography as a rhetorical form.

Second, Mitchell's analysis conjures up many of the same issues that one encounters in the academic debate over the possibilities of an environmentally responsible representation. On the one hand of this debate, we have advocates of scientific or literary realism who believe that with the proper care there can be a more or less mimetic relationship between signs and "natural" referents. On the other hand, we have social constructionists who believe in the inevitably arbitrary, cultural, and constructed character of signs and sign systems, however carefully crafted or functional. Each side distrusts the other: social constructionists believe that realists are naïve, and that they obscure the power relations invested in representation; realists believe social constructionists are solipsistic, excessively pragmatic, and anthropocentric to the point of denying (or at least being unable to acknowledge or perceive) the autonomy of the real.

One manifestation of this conflict emerges with the publication in 1995 of William Cronon's *Uncommon Ground: Toward Reinventing Nature*. Devoted in many ways to questioning our ability to adequately or objectively perceive, conceptualize, and represent nature, the collection was met with the counterpublication of Michael Soule and Gary Lease's 1995 *Reinventing Nature?: Responses to Postmodern Deconstruction*. The use of the interrogative (not to mention the subtitle's labeling of the opposition) carries with it accusations of hubris, the fear of ivory tower political paralysis, and, interestingly enough, a very rhetorical concern over *Uncommon Ground*'s ideas being co-opted for the cause of anti-environmentalism.[1]

From within the context of this particular intellectual conflict, then, representation presents us with a double bind: two options, both of which can result in an undesirable "capture" of the world. We either allow the world to be captured by the discourse of realism, which asserts the competence of representational mimesis to reproduce the world in words, or we allow it to be captured by the discourse of textuality, which claims that the interests of a natural Other are inevitably and utterly invaded by our own cultural baggage.

Yet despite—perhaps because of—the intractability implied by the books' titles (and the second edition of *Uncommon Ground* came with a revealing revision of the subtitle: *Rethinking the Human Place in Nature*), the true promise of their discussion lies in the very tension that makes such a call-and-response necessary: the productive power that simultaneously demands that we both reveal our understandings of nature to be cultural, historical, contingent, and ideological, on the one hand, and profess an obligation to the material world regardless—even if we can only reach it by proxy of our own inherited categories—on the other.[2]

In attempting to use that tension to steer my investigation, I find myself in a situation similar to Mitchell's. As a rhetorician, I am professionally sympathetic with the position of the social constructionists for the simple reason that I take it as my franchise to look at the way language does not simply transmit but rather constitutes understanding by establishing the parameters of intelligibility. However, as an environmentalist (and as a rhetorician concerned with how my own professional discourse is run through with power), I find myself similarly sympathetic to the realists, because their perspective asks a thorny question for which I have no easy answer: How can we discern a ground for environmental ethics within a framework of social construction when the material world then seems to be something always on the other side of the very concept of "Nature"—an Other without a voice (or at least without a voice we can understand except on our own terms)?

In what follows I argue that photography as a medium can supply the ground for a representational ethics that resists the very *possibility* of a complete capture of the natural, even if it does not imply complete success (some specter of absolute freedom) or an unobstructed view. What if, despite the apparent mutual exclusivity of realism and social construction, environmental photography always thoroughly partakes of both? To echo Mitchell, what if the only adequate formulation of the relation between photography, the material world, realism, and social construction was a paradox: Environmental photography is both thoroughly realistic and thoroughly socially constructed, and the thoroughness of each trait prevents the other from claiming a complete assimilation of the form?

Mitchell takes his notion of the photographic paradox from Roland Barthes, whose article "The Photographic Message" asserts that in photography we have "the co-existence of two messages, the one without a code (the photographic analogue), the other with a code (the 'art,' or the treatment, or the 'writing,' or the rhetoric, of the photograph)" (1977, 19). Barthes's work provides a useful set of concepts for the analysis of photography as a rhetorical form, and further, his change in focus from "The Photographic Message" to his later work, *Camera*

Lucida: Reflections on Photography (1981), demonstrates quite explicitly the same movement toward tension that I have identified in Mitchell and the debate over environmental representation and rhetoric: there is an initial emphasis on the cultural aspect of the photographic message, followed by discomfort with the limited possibilities of how photography can exist as a rhetorical form as a result, and an ultimate consideration of what is at stake in the resistance of a superstitious, realist, magic notion of photography. Ultimately, I would like to outline the possibilities of such a photography that is resistant to capture and to demonstrate how some of those possibilities are dramatized in Gary Paul Nabhan and Mark Klett's 1994 book *Desert Legends: Re-Storying the Sonoran Borderlands.*

In his earlier treatment of photography in "The Photographic Message," Barthes begins by pointing to what sets photography apart, "by definition," from other representational forms: its seeming transmission of reality itself. He admits to a "reduction" of the object in its photographic image, but not a *"transformation."* Because photography does not disassemble and reconstitute its object in the form of signs "substantially different from the object they communicate" (a code), we see it as pure denotation, a *"message without a code"* (1977, 17, emphasis in original). He suggests that all other representations can also *seem* to be messages without a code, but their analogical messages are always superceded by secondary coded ones. Because they are constituted by the codes of their given medium (brushstrokes, shaped material, pencil scratches), all other representations carry as their message the signifier of style: this or that painting, sculpture, or drawing is not "real"; it is "realistic." Their analogical quality is a generic choice, not an ontological necessity; the objectivity of these forms is "read as the very sign of objectivity" (1977, 18).

Photography's uncoded analogical message, however, appears so exact as to "fill up the photograph and leave no space for a second-order message" (ibid.). This "feeling [. . .] of analogical plenitude" explains why, for example, the adequate description of a photograph seems an impossible task:

> [*T*]*o describe* consists precisely in joining to the denoted message a relay or second-order message derived from a code which is that of language and constituting in relation to the photographic analogue, however much care one takes to be exact, a connotation: to describe is thus not simply to be imprecise or incomplete, it is to change structures, to signify something different to what is shown. (1977, 18–19, emphasis in original)

Yet immediately after articulating this view of photography as purely analogical, Barthes rejects it as having "every chance of being mythical" (1977, 19). Far from being exhausted by and filled in with denotation, the form most definitely takes on secondary, coded messages in its production, dissemination, and reception. For example, photographs are taken from particular angles and are framed or cropped to include or exclude certain objects. Often one particular shot (composed differently from its peers) is chosen out of many for inclusion either in newspapers, magazines, books, or exhibitions, and those forms of dissemination provide context (linguistic or otherwise) for its potential message. Ultimately, photography is not only perceived but, as Barthes points out, "*read*, connected more or less consciously by the public that consumes it to a traditional stock of signs" (ibid., emphasis in original). If there is a paradox of photography, it is not the coexistence of denotative (analogical) and connotative (signifying) messages—such a coexistence is found elsewhere as well. Rather, what is unusual is the manner in which we, as viewers of photography, inevitably impose the latter on the former: a code emerges out of its own absence.

Though such an analysis surely establishes a special status for photography's noncoded denotation, Barthes defers further investigation into the nature of that message: "Since the denoted message in the photograph is absolutely analogical, which is to say *continuous*, outside of any recourse to a code, there is no need to look for the signifying units of the first-order message" (20, emphasis in original). Instead, he devotes the remainder of his article to "forecast[ing] the main planes of analysis of photographic connotation" (ibid.), with the ultimate conclusion that the importance of doing so rests in the way connotation "makes of an inert object a language and [. . .] transforms the unculture of a 'mechanical' art into the most social of institutions" (1977, 31). The primary focus of analysis is placed on the way photography's coded message takes strength by obscuring its own coded nature.

This emphasis on the investment of social meaning into all representations is a perspective with which rhetoricians (specifically rhetoricians interested in environmental discourse) are very familiar. It demands attention to the possibility that ideas of "the natural," if seen as objective, neutral, or transparently representational, foreclose on debate (one cannot argue with "the way things are," with "what is natural," or with "the facts"). It also takes as a given the "typical process of the naturalization of the cultural" (26).[3] To be sure, an awareness of connotation, and the way understandings of nature (inevitably cultural) constrain and enable the possibilities of environmental behavior, is essential.

Yet, as I said before, there is a shift in attitude from "The Photographic Message" to *Camera Lucida*. Whereas in the former there seems to be some sort of approval or at least positive fascination with the way in which man "makes of an inert object a language and [. . .] transforms the unculture of a 'mechanical' art into the most social of institutions," in the latter Barthes speaks of how, "looking at certain photographs, [he] wanted to be a primitive, without culture" (1981, 7). He proceeds to describe the conflict not in terms that seem to imply the quickening of something lifeless, but rather the deadening of something already culturally alive:

> [T]his disorder and this dilemma, revealed by my desire to write on Photography, corresponded to a discomfort I had always suffered from: the uneasiness of being a subject torn between languages, one expressive, the other critical; and at the heart of this critical language, between several discourses, those of sociology, of semiology, and of psychoanalysis—but that, by ultimate dissatisfaction with all of them, I was bearing witness to the only sure thing that was in me (however naïve it might be): a desperate resistance to any reductive system. For each time, having resorted to any such language to whatever degree, each time I felt it hardening and thereby tending to reduction and reprimand, I would gently leave it and seek elsewhere: I began to speak differently. (1981, 8)

Barthes indeed speaks differently in *Camera Lucida*. While not refuting his earlier formulation of photographic signification as both denotation and connotation, and while still viewing the denoted message of photography as continuous, and therefore not constituted by "signifying units," he nevertheless finds that there is room for and value in investigating the rhetoric of the message without a code, for delineating not how it functions as a simple vehicle for connotation, but rather how it can exist for spectators as a magic, and he determines that this is both the genius of photography and the locus of effects that determine which photographs matter to him. Certain photographs, he claims, exert an attraction based upon their ability to move a viewer:

> [S]uddenly a specific photograph reaches me; it animates me, and I animate it. So that is how I must name the attraction which makes it exist: an *animation*. The photograph itself is in no way animated (I do not believe in "lifelike" photographs), but it animates me: this is what creates every adventure. (1981, 20)

The attraction of photography, then, is hinged upon an effect we could consider rhetorical. Yet the message that affects the viewer is constituted not through connotation, through a coded signification that the viewer deciphers, but rather through the reception of the photograph as something purely and perfectly analogical, through the belief that it comes unburdened by a code. And in this Barthes moves from a consideration of how photographs are culturally invested sign-forms to how they are effective when taken with utter belief: "I wanted to explore it not as a question (a theme) but as a wound" (1981, 21).

Instead of falling back solely to the terms denotation and connotation, Barthes supplements them with the slightly different *punctum* and *studium*. The *studium* is the connotational aspect of the photograph, that which we cooly grasp as the result of "a classical body of information" (25–26) and which emotionally involves us only through "the rational intermediary of an ethical or political culture" (26). For example, when I view Diane Arbus's *Child with a Toy Hand Grenade in Central Park, N.Y.C. 1962*, the photograph's *studium* is that which prompts me to reflect on the nascent militarism signified by the toy grenade the boy is holding and the inner tension conveyed by his posture and facial expression. It is what cues thoughts concerning the cold war history that gives the image its contextual meaning and the turbulent decade to come that it foreshadows. Similarly, a photograph such as Ansel Adams's *The Tetons and the Snake River, Grand Teton National Park, Wyoming, 1942* invokes inherited notions of the American West, the atmosphere of the nineteenth-century American sublime, and also—by virtue of both its similarity to and difference from Thomas Cole's painting *View from Mount Holyoke, Northampton, Massachusetts, after a Thunderstorm (The Oxbow)*—the tropes of the Hudson River School.

In contrast, when we turn to Barthes's *punctum*, it becomes possible to perceive the wounding power of photography:

> [I]t is not I who seek it out (as I invest the field of the *studium* with my sovereign consciousness), it is this element which rises from the scene, shoots out of it like an arrow, and pierces me. [. . .] A photograph's *punctum* is that accident which pricks me (but also bruises me, is poignant to me). (1981, 26–27)

Specifically insightful in this formulation of the *punctum* is the way in which the movement of influence from viewer to the object of a photograph is reversed. The *studium* is the invasion of photography by language that Burgin asserted was absolute, yet the *punctum* seems to originate in the photographic object itself, as a disruption of that invasion, a counterstrike. The *punctum* posits a resistance and an imaginary counteragent in the power relation of photographic representation.

One form of the *punctum* is the detail that seems to resist the photographer's intentions. Barthes sees a photo of a boy in Little Italy in 1954—a very real-looking gun held to his head as he smiles at the camera—and comments on the unintended: "What I stubbornly see are one boy's bad teeth" (1981, 45). Another form of the *punctum* bears on time and death. Looking at a photograph of would-be assassin Lewis Payne, waiting in his jail cell to be hanged, Barthes notes that the "photograph is handsome, as is the boy: that is the *studium*. But the *punctum* is: *he is going to die*. I read at the same time: *This will be* and *this has been*; I observe with horror an anterior future of which death is the stake" (96, emphases in original). What makes both of these forms of the *punctum* possible is a perspective that admits of an exceptional ontology of photography within the range of representational forms. Within this perspective, photography is completely contingent, irreducible to generalities, ideal forms, or tokens. In its exact replication, it is a magic of sorts, "literally an emanation of the referent" (80), and as such, its representation is in some final and absolute way outside of the photographer's control and affective in its sheer "evidential force" (89). Despite the most conscientious "composing" of a shot, there is an independent material object, of which the photograph is a trace, with which we have to make our meanings.

What, then, are the specific effects of this view of photography, and how do they relate to the problem of environmental representation being complicit in the dual capture of nature by realism and textuality? On the one hand, as a result of seeing photography as an exceptional, uncoded denotation, the *punctum* resists textuality's assumption that objects of representation are invariably, completely, and irretrievably invaded and obscured by cultural signification. While recognizing the *studium* is an aspect of accepting and moving with the photographer's intentions—Barthes refers to it as "a contract arrived at between creators and consumers," a form of cultural "politeness" (28)—the *punctum* is never read intentionally; it is a result at most of the simple presence of the photographer. The *punctum* is what cannot be avoided; it is what the photographer "could not *not* photograph" (47), the wounding detail that comes as an inalienable part of the whole. While it is something that the spectator brings in a sense to the photograph (we come to the photograph with a certain potential for being wounded, pricked; we offer up a pensiveness that gives the *punctum* an opportunity to

emerge, to punctuate the experience of viewing, and to injure us), it
is "what is nonetheless already there" (55), the result of photography's
referent being necessarily real:

> Photography's Referent is [. . .] not the *optionally* real thing to
> which an image or a sign refers but the *necessarily* real thing which
> has been placed before the lens, without which there would be
> no photograph. Painting can feign reality without having seen it.
> Discourse combines signs which have referents, of course, but
> these referents can be and are most often "chimeras." Contrary
> to these imitations, in Photography I can never deny that *the
> thing has been there.* (1981, 76, emphases in original)

Thus photography resists textuality by virtue of that which defines it
in opposition to other representational forms: it is the "unculture of
a 'mechanical' art," an undeniable testimony to the existence of the
photographic object.

Yet, on the other hand, a notion of magic photography similarly
undercuts the claims of realism, in that its complete contingency and evi-
dentiary power cannot truly "say" anything about its objects. It attests to
only existence, not meaning: the resistance of the *punctum* to the cultural
investment of the *studium* is a resistance to any effort to make the object
of photography speak in coherent, illustrative terms. What we see in a
photograph "has been here, and yet immediately separated; it has been
absolutely, irrefutably present, and yet already deferred" (1981, 77), and
as such, "it cannot *say* what it lets us see" (100, emphasis in original).
If one were to encounter a photograph of a creature or an environment,
then one would be limited to the following: "I exhaust myself realizing
that *this-has-been*; [. . .] it is in proportion to its certainty that I can say
nothing about this photograph" (107, emphasis in original).

Of what use, however, could the resulting message of *environmental*
photography be? When the denoted message of the photograph claims a
mitigated autonomy from connotation by virtue of its lack of a code, and
when that message then seems to be capable of nothing more than an
authentication of existence, what kind of desirable environmental understanding is possible? The answer, as far as I can see, must be to admit
that an understanding of the natural is not what photography offers us,
at least not an understanding more pragmatically useful than that offered
by other forms. What the photographic message does offers us, in spite
of its limited scope, is a form of environmental motivation: animation.

For example, a magic environmental photography could unintentionally conjure up memories that deal with past environments in their

specificity. Barthes notes that "the *punctum* has, more or less potentially, a power of expansion [. . .] often metonymic" (45), and he describes his experience with such a capacity thus:

> There is a photograph by Kertész (1921) which shows a blind gypsy violinist being led by a boy; now what I see, by means of this "thinking eye" which makes me add something to the photograph, is the dirt road; its texture gives me the certainty of being in Central Europe; I perceive the referent (here, the photograph really transcends itself: is this not the sole proof of its art? To annihilate itself as *medium*, to be no longer a sign but the thing itself?), I recognize, with my whole body, the straggling villages I passed through on my long-ago travels in Hungary and Rumania. (1981, 45, emphasis in original)

The unintentional and indelible detail, referential only in the photograph, overtakes the centrality of the *studium* by evoking a visceral response anchored not in a traditional stock of intelligible signs but rather in the experiential memory of the spectator. This is not, of course, to say that one escapes culture completely in such a moment—memory is hardly an unproblematic immediacy—yet experiencing the *punctum* as a metonymic expansion does resist the cultural imperative that would make some aspect of the photographic object into merely the token of a recognizable type; instead, it refers the spectator back to the relative contingency of his or her own experience.

The *punctum* also has the ability to animate the spectator through the establishment of a "blind field" (1981, 57), a plane external to the photograph in which the object's existence continues. If photography that does wound the spectator is simply a frozen representation, then the animation of the *punctum* has the potential, on the contrary, to create "a subtle *beyond*—as if the image launched desire beyond what it permits us to see [. . .] toward the absolute excellence of a being, body and soul together" (59, emphasis in original) and thus engenders a concern for and interest in the photographic object, breaking the indifference with which we commonly experience photography in a world suffused with images of all kinds. By animating the spectator in this way—despite its inability to transmit some sort of transparent understanding of the "natural" objects it represents—environmental photography can contribute to the cause of environmental advocacy while still holding out the possibility of a partial autonomy of the real.

Nabhan and Klett's 1994 book *Desert Legends: Re-Storying the Sonoran Borderlands* provides an example of what an environmental photography both thoroughly cultural and yet resistant to complete cultural invasion might look like. I should make clear at the outset that *Desert Legends* does not set itself apart in this way through any structural difference in Klett's photography itself. The basic understanding of photography that I have laid out according to Barthes's notions of denotation, connotation, messages with and without a code, and the *punctum/studium* distinction holds for photography in general. I am tempted to say that in this way, photography is photography. Yet in the same way that an acceptance of Burgin's characterization of photography limits the intelligible messages it can take on as a representational form, an acceptance of Nabhan's more wide-ranging characterization of how cultural forms can take stock of nature expands its rhetorical potential beyond a simple "culture *or* nature" (or "textualism *or* realism") split. When a reader views Klett's photography in the context of Nabhan's accompanying text, it becomes possible to see it as both an inevitably coded (or culturally invested) form, and one whose vitality, at the same time, comes from recognizing that this coding or investment does not amount to a complete absorption of nature by culture.

In his emphasis on the cultural aspect of environmental knowledge and representation, Nabhan is in the company of a number of other environmental thinkers for whom the idea of nature as an entity distinct from culture has proven a disappointment.[4] This is partially the result of a pragmatic concern that to think of nature in that way precludes a consideration of how to responsibly live in the world: if nature is only nature when humans are absent from it, then responsible human habitation and land use are lost causes. On the contrary, his environmental strategy insists that we recognize the imbrication of human influence and the natural world. He argues that it is primarily through long-term and close relationships with landscapes that we can learn to act responsibly toward them, and that the lasting way of disseminating and maintaining such knowledge is through traditional stories and traditional practices. As such, he advocates understanding our knowledge of and interaction with the real as cultural.

Additionally, though labeling Nabhan a devotee of a strong version of social construction would be inaccurate, his cultural emphasis also is the result of a certain level of distrust with the claims of objectivity and universality that are the distinguishing marks of modern science. Nabhan talks about scientific culture as just that: *a culture*. So while he *is* a scientist, an ethnobotanist, he realizes that his particular field has the advantage of being "not just a science. It's a reservoir of stories" (1994, 178–79). Mitigating this view, however, and admitting that culture does not offer a clear view of the nature it interacts with, he notes that the desert dwellers he chronicles in the book,

[r]ather than singing in perfect pitch with other desert voices, [. . .] still sound like a *norteño* band warming up for an all-night dance. They just keep on tuning up their instruments and improvising, as if no one is listening or watching. But when they stop for a moment, they may catch the reverberations of their own echoing voices or look up and see their own reflections, like mirages in the desert sky. (1994, 6)

The "nature" of Nabhan's book, then, is one that we are always struggling and never quite succeeding to hear or understand, with the result that "[w]hile getting to know the desert intimately, [. . .] rural dwellers have left their share of *disastres, lagrimas,* and half-baked schemes all over the land" (1994, 5).

Despite the limitation of being able to confront nothing but our own echoes and reflections in the world, Nabhan offers up a prospect very similar to Barthes's in regard to encountering something that those legends we create cannot completely obscure:

I can only warn you that you'll be facing some thorny images, ones that may poke at you, puncture your armor, or fester within like so many cactus spines. [. . .] One of these days, when you least expect it, that festering thorn buried in your side will rise from your flesh to let you know that somehow, this place has become part of your inner workings. It may have been beneath your notice, but all the while it has been quietly leaving its mark. (1994, 6)

We can of course read this thorn or cactus spine as a cultural one, and we would be right to do so, but we also can allow a magic view of the "thorny images" (in this case I will use the images supplied by the book's photographer, Mark Klett) to constitute a potential for a poking, puncturing, and festering resistance to forgetting the ultimate origin of environmental culture in our interactions with the real. Allowing Nabhan's ideas about culture and nature to establish the parameters for our analysis of Klett's photographs, the images become something that Burgin's formulation could never allow.

When I first look at Klett's *Night-Blooming Cereus Hiding in Creosote Bush Nurse Plant,* the cultural nature of the photograph is apparent. Far from the picture representing an immediate "window" onto its object, the focal range necessary to sufficiently highlight the cereus and the differentiated border exposure make obvious the mechanical production of the photo. With Barthes, we could, through the "rational intermediary of an ethical and political culture," say that the cereus's fragility is on

Figure 12.1. Mark Klett, *Night-Blooming Cereus Hiding in Creosote Bush Nurse Plant*

display, and that the need for a positive human interaction with the plant is thereby implied (this need being the focus of Nabhan's text of the second chapter, "Cryptic Cacti on the Borderline," which is devoted to its plight). These are, of course, intentional meanings to be extracted from the photograph, having already been imposed by its construction within the milieu of late twentieth-century American environmentalism.

Yet despite the apparent capture of the cereus as a sign within that discourse, it is also, as the object of a photograph, a trace of the necessarily real, particular cereus that Klett photographed, somehow inaccessibly prior to that signification: "*This-has-been.*" We have no recourse to saying what that prior real thing means or meant, unless we reinsert it into a discursive formation that can assign it a meaningful place. However, as a photographic denotation, this cereus nevertheless escapes in a mitigated fashion the fate that would be suffered by a painted cereus, which of course never existed in absolute contingency to begin with. It is interesting to note, though, that the animation I feel upon looking at this photograph is a result of an oppositional yet somehow a cooperative relationship between the blind field engendered by feeling "*This-has-been*" and the *studium*'s connotation of environmental scarcity and the possibility of extinction. Both constitute different, somewhat contradictory, possibilities for how photography's message can exist for me. Together they form a powerful rhetoric that never settles on fixable meaning yet does not result in the paralysis of environmental concern.

The border exposure of *Discarded Target, White Tank Mountains* functions the same way it does in *Night-Blooming Cereus*; I cannot say anything so simple as "this is the way it is or was," because traces of a production process remind me that this is just that: a human production. The *studium* in this case gives me more trouble, because the image does not lend itself easily to any formulaic environmental conclusion. Is the doll-head target supposed to be a sign for the heavy-handed treatment of nature by man (as a dump, as something to be covered over by one of our "superior" products), or the resilience of nature (it grows out from under the target)? The photo follows chapter 3, "Hanging Out

Figure 12.2. Mark Klett, *Discarded Target, White Tank Mountains*

the Dirty Laundry," in which Nabhan argues for "celebrating that which is rotten" (63), and returning ref-use to use: organic materials to the ground, metals melted down for further reforging, and so on. And yet the target is not serving a second purpose: its tenure was exhausted when it acquired too many bullet holes. This photograph's *studium* (at least insofar as I recognize it from a cursory glance) disrupts the final satisfaction of the preceding chapter by frustrating any sense of resolution. It resists the text, then, but I still do not feel it as a *punctum*. That finally comes when I realize that I cannot get away from a desire to register the texture of the plant. Brittle? Surprisingly elastic? Similar to the textures and feelings of the plants I remember from hiking on the Sierra Nevada foothill trails just north of Reno, Nevada? Following Barthes, I feel myself there, and I can hardly stand the frustration of my desire to reach through the paper.

This interplay between a *studium*-based, or culturally mediated perception of a photograph, and one that proceeds from the *punctum* and photography's message without a code, should not be seen as a possibility limited only to *Desert Legends*, or even to photographs that find themselves anchored to a discussion of culture and nature, such as Nabhan provides. One could just as easily—once such an understanding of photography's dual message has at least been put forward as a possibility for the viewer—find them in photography as seemingly realist as that taken for scientific documentation, or as seemingly textualist as that which is embedded in a value-laden ecotourism pamphlet, where the nature depicted stands in as a sign of both far-off exoticism and the promise of legitimizing a "nature-loving" identity. The dual-message perspective offers a way out of the limited possibilities that pure realism or pure textualism offers us. And by offering a way out of the ostensible "purity" of either representational theory, it also offers us a way to minimize the problematic capture of the real by either side. This is a possibility that has been, far too often, obscured by an almost partisan commitment to arguing solely for one side or the other.

It is true that, ultimately, from the perspective of environmental *action*, one must admit that cultural meaning and categories are the human tools that allow us to take substantive, positive steps toward change. We can only act based upon our understanding of what nature is and how we might protect it—however limited, constructed, imaginary, and flawed that understanding might be. A resistance that reminds us of our limitations, however—in a way that surrender to an absolute capture of nature by culture never can—works to open up a small, yet radical, place where the real is an animating force that is neither irrelevant nor reducible.

Notes

1. In particular, William Cronon's (1995) "The Trouble with Wilderness; Or, Getting Back to the Wrong Nature," in *Uncommon Ground*, and Donna Haraway's (1991) work, *Simians, Cyborgs, and Women*. For reactions to Cronon, see *Environmental History* 1 (1996): 29–55. For objections to Haraway, see the Preface to Soule and Lease (1995), *Reinventing Nature*.

2. Readers familiar with this debate will, of course, note that these two books were published roughly at the same time, and that the subtitle "Reinventing Nature" was first used in Donna Haraway's 1991 book. However, I refer to *Uncommon Ground* it in its stead because Haraway was included in the book and took part in the UC-Davis seminar that spawned it, and because it seems a common dismissive move to emphasize Haraway's admittedly more provocative stance over other scholars who share a commitment to varying degrees of a social construction perspective (including Katherine N. Hayles, who appears in both *Uncommon Ground* and *Reinventing Nature?*).

3. I am here thinking of the antifoundational move of Judith Butler's (1991) "Contingent Foundations: Feminism and the Question of 'Postmodernism,' " her radical extension of social construction as materialization, in *Bodies That Matter* (1993), and Kevin Michael DeLuca's application of those ideas to environmental rhetoric in *Image Politics: The New Rhetoric of Environmental Activism* (1999).

4. This perspective is particularly germane to environmental writers committed to sustainable agriculture, such as Wes Jackson, Gene Logsdon, and Wendell Berry (whose work makes an explicit shift away from human absence as an ideal in "Notes from an Absence and a Return," from his 1972 essay collection *A Continuous Harmony: Notes Cultural & Agricultural*).

Works Cited

Adams, Ansel. 1999. *The Tetons and the Snake River, Grand Teton National Park, Wyoming, 1942.* Collection Center for Creative Photography, University of Arizona, Tucson. In *Praise of Nature: Ansel Adams and Photographers of the American West*, ed. Alexander Lee Nyerges, 106. Dayton, OH: Dayton Art Institute.

Arbus, Diane. 2003. *Child with a Toy Hand Grenade in Central Park, N.Y.C. 1962.* San Francisco Museum of Modern Art, San Francisco. In *Diane Arbus: Revelations*, ed. Diane Arbus, 104–05. New York: Random House.

Barthes, Roland. 1977. "The Photographic Message." In *Image, Music, Text*, trans. Stephen Heath, 15–31. New York: Hill and Wang.

———. 1981. *Camera Lucida: Reflections on Photography.* Trans. Richard Howard. New York: Hill and Wang.

Berry, Wendell. 1972. "Notes from an Absence and a Return." In *A Continuous Harmony: Essays Cultural & Agricultural*, 35–54. New York: Harcourt Brace & Company.

Butler, Judith. 1991. "Contingent Foundations: Feminism and the Question of 'Postmodernism.' " *Praxis International* 11 (July):150–65.

———. 1993. *Bodies That Matter: On the Discursive Limits of "Sex."* New York: Routledge.

Cole, Thomas. 1994. *View from Mount Holyoke, Northampton, Massachusetts, after a Thunderstorm (The Oxbow).* Metropolitan Museum of Art, New York. In *Thomas Cole: Landscape into History,* ed., William H. Truettner and Alan Wallach, 75. New Haven, CT: Yale University Press.

Cronon, William, ed. 1995. *Uncommon Ground: Toward Reinventing Nature.* New York: W. W. Norton & Company.

DeLuca, Kevin Michael. 1999. *Image Politics: The New Rhetoric of Environmental Activism.* New York: Guilford Press.

Haraway, Donna J. 1991. *Simians, Cyborgs, and Women: The Reinvention of Nature.* New York: Routledge.

Klett, Mark. 1994a. *Discarded Target, White Tank Mountains.* In Nabhan and Klett, 1994, 24.

———. 1994b. *Night-Blooming Cereus Hiding in Creosote Bush Nurse Plant.* In Nabhan and Klett, 1994, 67.

Mitchell, W. J. T. 1994. "The Photographic Essay: Four Case Studies." In *Picture Theory: Essays on Verbal and Visual Representation,* 281–322. Chicago, IL: University of Chicago Press.

Nabhan, Gary Paul, and Mark Klett. 1994. *Desert Legends: Re-storying the Sonoran Borderlands.* New York: Henry Holt & Company.

Soule, Michael E., and Gary Lease, eds. 1995. *Reinventing Nature?: Responses to Postmodern Deconstruction.* Washington, DC: Island Press.

13

The Test of Time

McLuhan, Space, and the Rise of *Civilization*

Tom Tyler

Rhetoric . . . may be defined as the faculty of discovering the possible means of persuasion in reference to any subject whatever.

—Aristotle, *The Art of Rhetoric*

All media work us over completely. They are so pervasive in their personal, political, economic, aesthetic, psychological, moral, ethical, and social consequences that they leave no part of us untouched, unaffected, unaltered. The medium is the massage. Any understanding of social and cultural change is impossible without a knowledge of the way media work as environments.

—Marshall McLuhan, *The Medium Is the Massage*

The Second Coming of Saint Marshall

In 1965, Tom Wolfe wondered whether media theorist Marshall McLuhan might be "the most important thinker since Newton, Darwin, Freud, Einstein, and Pavlov" (Stearn 1967, 15). George Steiner described him as "this enormously exciting iconoclast" (236). Even McLuhan's detractors, and there were many, could grant the value of his provocative interventions: despite considerable reservations, Raymond Williams described *The Gutenberg Galaxy* as "a wholly indispensable" book, and Jonathan Miller was willing to concede that "enough of the doors that he opens are exciting and productive to make him worth studying" (189, 236).[1] At the height of his fame, or perhaps infamy, McLuhan

was declared "the oracle of the electric age" by *Life* magazine, appeared
on the cover of *Newsweek*, was interviewed by *Playboy*, and hosted an
NBC TV show entitled *This Is Marshall McLuhan*.[2] Then, following the
"McLuhanacy" of the 1960s and early 1970s, and his death in 1980,
McLuhan's work and ideas dropped beneath the critical and popular
radar: interest waned, and books went out of print. It was not until
the eve of the new millennium that "another quasi-global outpouring
of interest and influence tied once again to emerging communications
technologies and information systems" developed, and Gary Genosko
could proclaim that "For better and for worse, a McLuhan renaissance
is in full swing" (Genosko 1999, 1). A raft of new publications on
McLuhan coincided with reprints and critical editions of his work,[3] while
Wired magazine verified this "second coming" (ibid.) by canonizing
McLuhan as its "patron saint."[4]

In this chapter I examine whether any of those exciting and productive doors that McLuhan left open might lead to theoretical resources
that could usefully inform the consideration of visual and environmental
rhetoric. In particular, I will take up McLuhan's notions of visual and
acoustic space and consider their utility for thinking about certain aspects
of a relatively new medium whose increasing importance and popularity
have largely overlapped with the resurgence of interest in the work of
Saint Marshall: that of digital games. I would like to investigate what
McLuhan's iconoclastic ideas can tell us if we consider at one and the
same time the complicated visual nature of digital games and the implicit
environmental discourses they comprise. Genosko suggests that "the
McLuhan legacy was singularly devoid of progressive political ideas and
remains largely the same today, with a few exceptions" (1999, 12). This
chapter might be considered, I hope, one of those exceptions. My objective
is to reflect on what McLuhan has to say about media as environments
in order to better understand the nature of the rhetoric, specifically the
environmental rhetoric, constituted by the medium of digital games. My
suggestion is that digital games, as a distinct medium, work to persuade
players of their immediate and continuing participation in the environment of which they are a part. By way of illustrative example, but also
in order to complicate the chapter's own message somewhat, I examine
Sid Meier's highly successful and enduring game *Civilization*.

McLuhan was interested in media not as channels of communication, not as vehicles for conveying a meaning or an idea from author
to audience or from producer to consumer, but as environments. A
medium will have an effect on us in virtue of the particular material and
social changes that it engenders, irrespective of any individual messages
that it is used to transmit. The impact of the telephone has been far

more significant than any of the specific uses to which it has been put.[5] Media, which include for McLuhan an extensive range of technological innovations, touch us, affect us, and alter us. The *medium* is the message, or perhaps the "massage."[6] Thus in looking for the possible means of persuasion, for the ways that media work us over, McLuhan finds a rhetorical import and impact within the medium itself, rather than in the message or content it conveys. It is the pervasive, persuasive function of media to shape us—personally, politically, psychologically—by the environments they create. It is this rhetoric of the environment with which we will be concerned, so that we might examine the environmental rhetoric of digital games.[7] Before doing so, however, it will prove useful to consider McLuhan's account of the environments created by two of the most important media developments: literacy and print.

An Eye for an Ear[8]

McLuhan was by no means the first to suggest that the development of literacy had a profound effect on culture and society, but the importance he accorded this technological innovation and the particular approach he took to its impact were novel.[9] McLuhan argues that writing, as a medium, prompted the rise of civilization as we know it, and it did so because it shifted us from one environment into another. With the advent of literacy, we moved from what McLuhan calls acoustic space into visual space. Acoustic space is the environment in which we live insofar as it is accessed predominantly by our sense of hearing. It is characterized, McLuhan argues, by inclusivity and a lack of central focus. "We hear equally well from right or left, front or back, above or below. . . . We can shut out the visual field by simply closing our eyes, but we are always triggered to respond to sound" (1960, 41). As Wordsworth suggested, "We cannot bid the ear be still" (quoted in McLuhan and Fiore 1967, 44). Thus, "The ear favours no particular 'point of view.' We are enveloped by sound. It forms a seamless web around us" (111). Members of those cultures who were dependent on speech and hearing for the majority of their communications, which is to say those who existed prior to widespread literacy, were accustomed to living in acoustic space. "Primitive and pre-alphabet people integrate time and space as one and live in an acoustic, horizonless, boundless, olfactory space" (57). Within this "sphere without fixed boundaries" (1960, 41) they experienced a rich, compelling connectivity to the world. With no central focus to this experiential sphere, an interplay between all of the senses remained possible, a "tactile synaesthesia"

(1962b, 17). "Acoustic" space, then, entailed a kind of total inclusivity, an immersion within the environment.[10] This unity was unavoidably lost with the arrival of literacy.

McLuhan argues that this technological innovation, particularly the phonetic alphabet, caused a shift in the relative significance of the senses. Writing promoted the eye rather than the ear. "Its use fostered and encouraged the habit of perceiving all environment in visual and spatial terms" (McLuhan and Fiore 1967, 44).[11] This emphasis on the visual, and the insistence of the distinct, uniform characters of an abstract, phonetic alphabet, brought with it a new kind of separation, of segmentation, unknown to the inhabitants of acoustic space. Modes of perception and patterns of thinking became increasingly linear, distinct, departmentalized. Print exacerbated and extended this predominance of the eye and the fragmentation it entailed.[12] The printing press, which McLuhan calls "a ditto device," provided "the first uniformly repeatable commodity, the first assembly-line, and the first mass-production" (1962b, 124; McLuhan and Fiore 1967, 50). This mechanization of writing continued the process of abstraction begun by the individually meaningless characters of the phonetic alphabet. Crucially, the foregrounding or isolation of the visual, and the consequent "separative and compartmentalizing or specialist outlook," tended to produce a "fixed point of view," on which "the triumphs and destructions of the Gutenberg era" would be made (McLuhan 1962b, 126–27).[13] McLuhan pins on printing a host of varied historical developments, including the rise of the nation-state: in rendering the vernacular visible and unified, print "created the uniform, centralizing forces of modern nationalism" (199).[14] This consolidation of the linear homogeneity of visual space was "the making of typographic man."[15] The culture of the West, the rise of this particular species of civilization, McLuhan suggests, was due in large part to the operation and effects of literacy: "Civilization is built on literacy because literacy is a uniform processing of a culture by a visual sense extended in space and time by the alphabet" (1964, 86).[16]

The Renaissance thus saw the creation of a new environment in which the visual ruled as tyrant. The "interplay of all the senses in haptic harmony" (McLuhan 1962b, 17), characteristic of oral culture and enduring into the age of the manuscript, was stamped out by the ditto device. Something of the seamless web of acoustic space was recaptured, however, with the onset of the "electronic revolution." Telephone, phonograph and radio all extended the oral and acoustic, of course, but so too, McLuhan argues, did a congregation of other technological innovations: "even our visual electronic forms, the telegraphic press, teletype, wirephoto, and TV are oral in character" (1958–1959, 169).

Like the immediate communication between members of oral culture, the instantaneity of electronic communication, now on a global scale, created an "auditory spatial structure [which] is a simultaneous field of relations," since "the oral is accidentally the spoken but essentially the instantaneous" (169). The successive, linear segmentation accentuated by the visual supremacy of typography gave way to an instant electric inclusivity in which all of the senses are involved, so that "[o]ur extended faculties and senses now constitute a single field of experience" (McLuhan 1962b, 5). With the electronic restoration of a tactile synesthesia, we become immersed within a new kind of acoustic space.[17]

The speed of communication in the electronic age renders the centralization characteristic of the mechanical age redundant. Information moves too fast for any center to keep up, ensuring that "[t]he new situation is not the old sponge pattern of intake from the margin and output from the centre, but of dialogue among centres" (1962a, 37). The "electronic conditions of implosion" thus create decentralized "centres-without-margins" (26, 23).[18] As McLuhan famously suggested, this electronic, dialogic interdependence "recreates the world in the image of a global village" (1962b, 31). The multiple centers are brought into direct contact with one another, so that everyone is involved and connected to everyone else. "We live in a single constricted space resonant with tribal drums" (ibid.). This newly recaptured "tribal web" (1964, 84) does not ensure cooperation or conformity, however. In fact, the conditions of the "tribal-global village" (Stearn 1967, 280) produce more discontinuity, diversity, and disagreement than the uniform consistency of visual space ever did.

War and Peace in the Global Village[19]

Digital games illustrate McLuhan's notion of a new acoustic space especially well and thereby provide an opportunity to explore the utility of his particular understanding of the rhetorical effects of new media.[20] Digital game genres are many and varied, but all such games create, as a result of their technological modus operandi, a distinct kind of environment that participating players must inhabit.[21] Immersed within this interactive game environment, a player's experience is qualitatively different, not just from reading a text or viewing an image but also from watching television or film.[22] By inviting, or, rather, requiring of the player, an immediate engagement with and connectivity to the medium, games are, in McLuhan's terms, more acoustic than they are visual. Online gaming provides perhaps the best, or most obvious, example of the inclusive, decentered nature of digital gaming. In role-playing games such as *World of WarCraft* and *Lineage,* or action games such as *Counter-Strike* and *Quake,* millions of players from across the globe meet within vividly realized virtual environments, communicating and interacting instantaneously. In order to illustrate fully the acoustic quality of digital games, however, I turn to an example from a rather different genre, a game that began life as an offline, single-player, turn-based affair, and that upon first appearances seems, therefore, to be an altogether more visual experience.

Based in part on an intricate 1980s board game of the same name, Sid Meier's *Civilization* was released by MicroProse in 1991. The computer game was initially only modestly successful but went on to secure a series of awards and increasing popularity. A new version, *Civilization II,* with tweaked gameplay and improved graphics, was released five years later, significantly increasing the number of players and prompting in turn two further expansion packs. *Civilization III* (2001) and *Civilization IV* (2005) followed, each time to widespread critical and popular acclaim within the games industry, making the series one of the most successful ever. The rise of *Civilization* has been unremitting and continues to this day.[23]

Players of *Civilization* start the game in control of a settler, the sole representative of their chosen tribe, in 4000 B.C. Their immediate goal is to found a city, but from here the ultimate objective over the centuries to come is to establish a globe-spanning civilization. Play unfolds on a huge world map, which is gradually revealed over the course of successive turns. The game belongs to, and indeed helped define, the now-popular "4X" strategy genre: exploration (of the world), expansion (of your empire), exploitation (of resources) and extermination (of rivals).[24] Players pursue their imperial agenda by founding additional cities, constructing military units, researching science and technology, building Wonders of the World, trading with competitors, and so on. By these means, players hope to "Build an empire to stand the test of time," as the game's original strapline urged (Figure 13.1).

Civilization, the game, seems to exhibit all of the hallmarks of a visual form according to McLuhan's schema, as his comments concerning civilization, the process, begin to suggest:

> A goose quill put an end to talk, abolished mystery, gave us enclosed space and towns, brought roads and armies and bureaucracies. It was the basic metaphor with which the cycle of civilization began, the step from the dark into the light of the mind. The hand that filled a paper built a city. (1969a, 14)

Figure 13.1. *Civilization IV* Map

Most importantly, the game plays out on a map, that supremely visual medium which reduces all the world to a homogeneous, geometric space and entails thereby a kind of sensory blindness (McLuhan 1962b, 11).[25] This playing area is, in effect, a sophisticated, animated board, not unlike those used for traditional board games, and the playing pieces are similarly restricted in their linear, sequential movement. The game even simulates line of sight for these units, a favored point of view that permits players to see only those parts of the board that have been visited, and only those enemy units that are close by (the so-called "fog of war"). Ted Friedman has argued that in simulation games such as *Civilization*, the map itself thus becomes the protagonist in a geographical narrative (Friedman 1999). Space is conceptualized by abstract images, and the games operate as maps in time, "dramas which teach us how to think about structures of spatial relationships" (ibid.). These structures are *visual* spatial relationships, of course, and the map's movement through time accords with the strict linearity described by McLuhan. The game progresses in rigidly sequential turns: players must await their go and must move each of their individual units one at a time, one after another. Even a tribe's scientific development proceeds in strict order: any given advance (wheel, gun powder, electricity, etc.) cannot be researched until prerequisite technologies have been mastered, and progress is represented by a branching, arborescent "technology tree."[26]

Further, *Civilization* has remained, at least until its recent incarnations, an essentially single-player experience. Multiplayer functionality has been available since *Civilization II*, but the measured, turn-based pace and extremely lengthy course of each game has, for the most part, made the involvement of several players impractical.[27] Just as, according to McLuhan, the alphabet and its mass-produced texts resulted in a largely solitary communicative experience, at least compared to the simultaneity of dialogic interchange, so *Civilization* gives rise to a predominantly isolated mode of play. Finally, in taking the role of a near-immortal sovereign, forging a homogeneous empire under their personal, direct command, players manage a highly centralized state.[28] Those nomadic settlers with which the game begins swiftly set down roots, founding a capital city complete with imperial palace. The tribe becomes a nation, unified under absolute, central control.

Civilization is altogether more acoustic than it would initially appear, however. First, there is for the player no favored point of view. Where first- or third-person action games such as *Counter-Strike* or *Tomb Raider* require the player to assume a single, fixed perspective, *Civilization* allows continual access to any part of the world that has been revealed. There is thus no central focus but rather equal perception in all directions.[29] Further, despite the fact that your first city is a nominal capital, the cities that you go on to found are no less important in terms of the resources they exploit and the improvements they build. In the early stages of the game they can remain relatively autonomous, self-contained centers in their own right, but once conjoined by road or rail, they form a network, a web even, of mutually supporting bases. This single space, resonant with the "tribal drums" of instantaneous communication, facilitates a dialogue among centers without margins.[30] In McLuhan's terms, then, the empire established by a tribe is not so much a centralized nation as a growing, meshlike global village.

Moreover, there is a good deal more to the playing area, the space of *Civilization*, than the margin-free world map. Images may well be, as Friedman suggests, the clearest way to represent visual space, but an aspiring imperator must command more than a scenic view. In addition to the map, players also have access to a large number of charts and tables that relate all manners of essential information. Each "city display" conveys statistics regarding population, food stores, trade and corruption, building projects and improvements, and so on. (Figure 13.2), while various advisors or "ministers" report on the nation's defenses, relations with foreign powers, the morale of the populace, the status of scientific research, and so on. Play does not stop at the map in time, then, but incorporates numerous charts in time (Friedman, 1999, n. 8), immersing the player in a multidimensional, acoustic experience. Friedman argues that digital games reorganize structures of perception by requiring players to internalize the logic of the program, to "think like a computer" (Friedman 1999). The immersive pleasures of an effective game of *Civilization* are due at least in part to the effortless, instantaneous way in which an experienced player will begin to access and manipulate the heterogeneous maps, charts, and tables.[31] Players come to identify not with the tribal leader they purportedly control but, rather, with the roles and functions carried out by the game itself. The player's perspective is not that of a God's-eye view, a prospect from above that surveys a uniform, geometric visual space, but rather of a computer, an exploration from within the sphere of instantaneous electronic communication. Just as there is no single geographical center to the player's expanding empire, so there is no single, privileged subject-position for the player who is dispersed across a varied and changeable acoustic environment.[32]

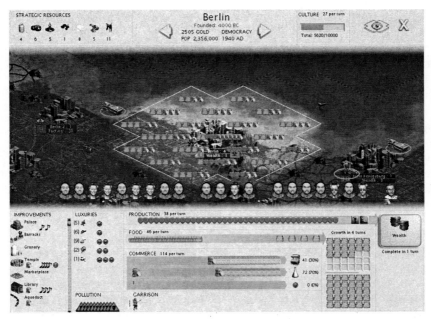

Figure 13.2. *Civilization III* City Display

McLuhan argues that the sheer volume and diversity of information with which consumers are faced in the electronic age necessitate a "producer-oriented" approach, whereby consumers become discriminating co-creators of their own media experiences (1958–1959, 167). The particular kind of interaction in which player and computer engage certainly requires that the former take a proactive part in the construction of the gaming experience. Immersion within the medium of the game ensures that the process of play is no hermeneutic matter of discerning *Civilization*'s "message." The environment, both geographic and otherwise, changes during the game as a direct result of the player's actions. Terrain is mined or irrigated, taxation is raised or reduced, and governments are replaced or restored. The game continually demands decisions from the player, at both the local and global levels, and every one will have consequences for the success of the tribe, sometimes within just a year or two of game time, sometimes decades or centuries later.

At the same time, David Myers has further argued that it is precisely the "transformative" aspect of *Civilization*'s gameplay that makes it such a compelling experience (2005). With each technological advance, with each change in government, and especially as the powerful Wonders of the World are secured by one tribe or another, the rules of the game are significantly transformed, requiring players to reassess their immediate goals and perhaps modify their style of play. A full understanding of the far-reaching effects of these transformations can only be gained by replaying the game and by experimenting with different strategies and tactics, many times over.[33] The recursive nature of replay, the experimental adaptation that comes with successive iterations, results in an acoustic space that is never consistent but always evolving.[34]

Players are submerged, then, in an environment that molds their actions, prompts their responses, and works them over: the medium is the massage. But at the same time, as we have seen, they work to shape their environment, actively modifying their surroundings, pursuing "the tilling of virtual landscape" in the broadest sense (Myers, 2005). Players are not subsumed into or absorbed by this virtual, acoustic environment, and thinking like a computer need not mean that they become one with the software: there is still a useful distinction to be drawn between the masseur and those massaged. Players of *Civilization* exist within the game environment, experiencing a rich, compelling connectivity to the world, but, engrossed as they may be, connection is not assimilation.[35] If the medium is the *message*, then the directive or intimation here is that the player is an active, involved part of the environment. Every decision will impact on the state of the game, provoking consequences that the player will in turn experience.[36]

Digital games, including the seemingly linear and visual *Civilization* series, provide an illuminating illustration of McLuhan's notion of a new, inclusive, acoustic space. Players are immersed in an engaging, centerless, electronic environment. Strictly speaking, if we are to follow McLuhan, then an individual game such as *Civilization* simply illustrates this environment: each time we take up Sid Meier's challenge, we invoke in miniature the wider acoustic space of which the medium as a whole is a part. Digital games in their entirety comprise one contributing component of the distinctive acoustic environment that has emerged since the nineteenth century, with the advent of telegraph, radio, television, and so on. McLuhan thus provides a productive probe with which to explore digital games, but, at the same time, a close examination of this particular electronic medium helps to clarify our understanding of the new acoustic space to which he draws our attention. The inclusivity and immersion that characterize acoustic space should not be conceived as an assimilation or absorption of individual subjects into that space. Notoriously compelling as *Civilization* is, and much as the game works us over personally, aesthetically, and even psychologically, the player, like the denizens of the wider acoustic space, remains an active participant within the environment.

Talk to the Media, not the Programmer[37]

In his analysis of *Civilization*, Myers's focus on the transformative, recursive nature of gameplay accords closely with McLuhan's insistence on the necessity of analyzing media as environments rather than as vehicles or channels of communication. Both writers distance themselves from textual analysis, concentrating instead on the internal mechanics and physical or psychological effects of their chosen media.[38] Myers goes so far as to suggest that, in general, game backstories, the carefully realized social and cultural milieus in which game events take place, "have no real relevance to computer gameplay" (2005).[39] What matters is "not the setting or the characters or the plot, but the *relationships* among the game's signs and symbols as adjudicated by the game rules" (2003a, 9). In fact, *Civilization*'s particular representation of world-historical development, its backstory in the broadest sense, has been the subject of much discussion and ideological critique. In this final section, I explore the explicit but equivocal environmental rhetoric of the game. Conflicting textual analyses seem to suggest, I argue, that it is important to attend here not just to the message but to the medium itself.

Civilization's rhetoric with regard to environmental issues is, on first sight, unpromising. Even setting aside the imperial and colonialist objectives that players are required to take up,[40] and the peculiarly occidental assumptions regarding cultural and governmental progress—factors that between them have borne the brunt of criticism from the game's always-respectful detractors (Stephenson 1999; Henthorne 2003; Poblocki 2002; Lammes 2003; Bitz 2002; Douglas 2002)[41]—*Civilization*'s approach to ecological questions remains problematic. At the start of play a tribe's nomadic settlers find themselves amidst virgin landscapes: lush plains and grasslands, dense forests and arid deserts, hills and mountains, swamps, jungles, and more. Once settled, however, and as the expanding tribe establishes one city after another, the terrain changes. The land around each settlement is developed: roads are laid to connect urban centers, fields are irrigated beside available waterways, and hills are mined. As technology advances, railroads begin to crisscross an intensively farmed landscape. Gradually but inexorably, the diverse natural environment is replaced by an increasingly homogeneous topography, a highly developed, uniformly industrial conurbation. Only by such expansion and cultivation can players hope to survive and succeed.

One might argue that there is an implicit environmental message here: *Civilization* draws the player's attention to the fact that increasing urban settlement *does* deplete and destroy the surrounding land. There is a lesson here, perhaps, that the demands of concentrated population growth directly counterpose environmental concerns. And, in fact, swelling populations and escalating production present players with a particular, new problem. The accumulating factories, offshore platforms, and manufacturing and power plants all increase the likelihood of environmental pollution. Represented as a death's head skull, or a garish blight on the land, pollution not only lowers a player's final score but nullifies the benefits that industrialization and intensive farming had initially bestowed. The problem of pollution persists as the game progresses and, if left unchecked, can lead to global warming, rising sea levels, and barren coastal farmlands. Only by constructing city improvements, such as solar and hydroelectric power plants, recycling centers, and mass transport systems, can the threat of pollution be averted. Nuclear power stations reduce the likelihood of immediate pollution but bring the possibility of a ruinous meltdown, and the deployment of nuclear missiles has similarly devastating effects. If there is a textual message to be detected within *Civilization*, then it is surely that willful disregard of the environmental consequences of one's actions results in dire costs and penalties.[42]

And yet pollution is just one obstacle among many that must be addressed by the player whose key objectives are, as we have seen, exploration, expansion, exploitation, and extermination. There is certainly no incentive to *reduce* the consumption of resources, or, indeed, to regard the game's geography as anything other than a storehouse of goods and material opportunities.[43] It is, in fact, a simple matter to "clean up" polluted terrain: teams of engineers or workers assigned to the task swiftly remove all evidence that a problem ever existed. Even nuclear fallout, from missile or meltdown, can be scrubbed from the map in a few game years.[44] Though repeated cleaning can become tedious in gameplay terms, the implication is that environmental pollutants, nuclear or otherwise, are an easy matter with which to deal. Indeed, along with civil unrest, corruption, waste, and a number of other elements considered detrimental to smooth gameplay, pollution was removed from *Civilization IV* in favor of a streamlined "city health" system, shifting the ecological focus from global environment to metropolitan well-being.

In *Ecospeak*, Killingsworth and Palmer define rhetoric as "the production and interpretation of signs and the use of logical, ethical, and emotional appeals in deliberations about public action" (1992, 1). Interpretation of the signs that are produced and manipulated within *Civilization* seems to yield an ambivalent message. Player deliberation about public action, about choices concerning development and growth within the game, will be informed by a rhetoric that seems both to acknowledge and to elide environmental concerns. If we are interested in the game's possible means of persuasion, in the ways in which it massages and works us over, however, we might do better to attend to the medium rather than the purported message. Over the decades and centuries, players experience, by their immediate and engaged immersion within the game environment, the direct and long-term consequences of their own actions. Not just the decisions they make concerning industrial growth and development, but every unit or building they construct, every adjustment they make to taxation or scientific research, and every trade or diplomatic negotiation they pursue, will impact on the way in which the game ultimately plays out. Players are participants, actively involved in the medium-as-environment, experiencing firsthand a compelling connectivity to the world. *This* is the key environmental rhetoric of the digital game *Civilization*, the message of the medium.

Raymond Williams suggested, or perhaps hoped, that the "particular rhetoric of McLuhan's theory of communication is unlikely to last long" (1974, 128). The McLuhanacy of the 1960s and 1970s did indeed pass, but as Gary Genosko observes, at the start of the new millennium we find ourselves immersed in a McLuhan renaissance. Is this renewed interest warranted? Does McLuhan's work speak to the electronic revolution he envisaged? The utility of McLuhan's provocative writings lies, I think, at least within the context of environmental rhetoric, with his insistence on the complicated relations that pertain between the visual and the acoustic. The temptation to sensory blindness, to conceiving digital games as a primarily visual medium—linear, uniform, characterized by sequential segmentation—is contested by the games' irresistibly acoustic qualities, their inclusive, immersive, decentered interactivity. In the sensory interplay of this digital, acoustic space, participants are engaged and implicated in the suasive rhetoric, the medium as environment, which touches, affects, and alters them. McLuhan's own rhetoric will more than likely stand the test of time, precisely because he left open exciting and productive doors and provided rich and largely untapped theoretical resources, through which we might yet address acoustic and visual questions of ecospeak and ecosee.

Notes

1. McLuhan's varied reception in the 1960s is captured well by three contemporary collections: Stearn, Rosenthal, and Crosby and Bond. For an excellent overview of McLuhan and Williams's contrasting approaches to the study of the media, see Lister et al. (2003, 72–92).

2. *Life*, February 25, 1966; *Newsweek*, March 6, 1967; McLuhan, 1969b; *This is Marshall McLuhan: The Medium Is the Massage*, NBC TV (March 1967), produced and directed by Ernest Pintoff and Guy Fraumeni; a scathing review of the NBC show appeared in the *New Yorker* (Rosenthal 1968, 82–87).

3. On renewed interest in McLuhan's work from writers such as Baudrillard, Virilio, Poster, Kroker, and De Kerckhove, see Lister et al. (2003, 73–74), Genosko (1999, 8–12), and Horrocks (2000, 14–18). Reprints have been issued by Gingko Press.

4. *Wired*, 4.01 (January 1996) http://www.wired.com/wired/archive/4.01/, accessed June 24, 2008.

5. See McLuhan (1964, 265–74) for his discussion of the telephone, and the Preface to the third printing (v–x) for a discussion of media as environments.

6. McLuhan developed a number of variations on this, his most famous sound bite: see Levinson (1999, 35–36). For more on media as environments see McLuhan (1969a, 30–31, 1966).

7. On "McLuhan as Rhetorical Theorist," see Gronbeck 1981; see also McLuhan's (2005) work *The Classical Trivium*.

8. McLuhan (1964, 81, 1967, 44).

9. In what follows, I am interesting in exploring the potential uses of McLuhan's characterizations of acoustic and visual space rather than assessing the validity of his claims regarding their correlation to distinct historical periods. On the impact of literacy, see Havelock (1963) and Ong (1982); for discussion of parallels between McLuhan and previous writers, see Duffy (1969, 26–31); for a critique of the determinacy that McLuhan accords literacy and print, see

Williams (in Stearn 1967, 186–89). Of particular interest within the context of the present chapter is David Abram's discussion of literacy, informed by Merleau-Ponty, which explicitly considers the debt owed to the "more-than-human" writing of the natural world, and whose approach to the relations that pertain between literacy, conceptions of space and time, and synaesthesia mirrors that of McLuhan (Abram 1997).

10. See McLuhan's (1960) "Acoustic Space" for his most sustained treatment of the matter. Duffy provides a concise and accessible explanation of the key characteristics (1969, 22–25).

11. There are distinct parallels here with the thinking of Benjamin Lee Whorf, whose work McLuhan had read. See Whorf (1956), and McLuhan et al. (1977, 182).

12. Note, however, that for McLuhan there are key parallels as well as differences between nonliterate and preprint literate cultures; see Duffy (1969, 23–25).

13. On the uniformity, repeatability, and "fixed point of view" engendered by print, see *The Gutenberg Galaxy* (1962b, 124–27). McLuhan's use of the latter term derives from Gyorgy Kepes's (1944) *The Language of Vision*.

14. "The mechanization of writing mechanized the visual-acoustic metaphor on which all civilization rests; it created the classroom and mass education, the modern press and telegraph. It was the original assembly-line. Gutenberg made all history available as classified data: the transportable book brought the world of the dead into the space of the gentleman's library; the telegraph brought the entire world of the living to the workman's breakfast table" (1969a, 15)

15. This phrase is the subtitle of McLuhan's (1962b) *The Gutenberg Galaxy*.

16. Genosko has discussed the unrepentant logocentrism of McLuhan's account of the transition from oral to literate culture (1999, 36–41).

17. Much has been made of the relevance of McLuhan's ideas on the electronic revolution to considerations of digital and online media; see, for instance, several of the essays in Strate and Wachtel (2006).

18. For an elaboration on this point, see "The Electronic Age—The Age of Implosion" (1962a, 23–27).

19. McLuhan and Fiore (1968).

20. I eschew the term *video*game for reasons that will become clear in a moment. McLuhan's work has rarely been applied to digital games; for exceptions, see David Miles's discussion of the "multimedia novel" *Myst* (Miles 1996); and Lister et al. (2003, 271).

21. Instructive parallels might be drawn here, if space allowed, between gaming environments considered from a McLuhanesque perspective and Huizinga's notion of the magic circle (Huizinga 1955, 10), the latter having been widely taken up within game studies (Salen and Zimmerman 2004, 94–98).

22. See, for instance, Aarseth's much-cited discussion of the "extranoematic" effort required by "ergodic" media (Aarseth 1997).

23. For a history and an informative account of *Civilization*, see Myers (2003a, 131–46).

24. The term is Alan Emrich's, quoted in Myers (2003a, 136). The *Civilization II* manual suggests an alternative configuration of "basic impulses"—exploration, economics, knowledge, conquest—which perhaps provides a better indication of the *variety* of the gameplay (Reynolds 1996, 2–3).

25. For a brief discussion of the map within McLuhan's work, and its relation to nation states, see Neve (2004).

26. In *Civilization III*, technological development is further restricted to successive historical ages (ancient, medieval, industrial, modern), although *Civilization IV* revoked this innovation and indeed made scientific development somewhat more flexible. All versions of the game include *alphabet* as an early technological advance, although, pace McLuhan, it is accorded no greater significance than any of the others.

27. *Civilization III* added a variety of features to facilitate multiplay: time-limited turns, simultaneous play, hotseat games, play-by-email, voice chat, and so on.

28. From *Civilization III* onward, these leaders are named and depicted as specific historical figures, such as Caesar, Gandhi, Bismarck, and so on.

29. It is worth noting too that there are no extreme eastern or western edges to the animated board: the world map wraps east to west. The 2-D playing area is not quite the acoustic sphere described by McLuhan, then, but it does begin to approach a marginless global space.

30. Admittedly, a nation's capital *is* geographically central, in two regards: *corruption* can increase in distant cities, and, in *Civilization III*, capitals play a small part in the impact of a nation's *culture* on its neighbors. Neither one amounts to the centralized "sponge pattern" of intake from the margin and output from the center described by McLuhan, however.

31. Friedman suggests that we might productively consider this collaboration between player and computer a kind of single "cyborg consciousness" in which the player becomes, on one level, "an extension of the computer's processes" (Friedman 1999).

32. Miklaucic accounts for *Civilization*'s heterogeneous, multidimensional charts and graphs, and their part in the player's dispersed and decentered engagement with the game, in terms of Bolter and Grusin's notion of hypermediacy; see Miklaucic (2003).

33. Myer's fascinating analysis in fact concerns the recursive, nonlinear, spiral-like trajectory of replay within *Civilization*. His account of the recursive development of successive editions of the game itself is similarly instructive.

34. For an illuminating account of iterative gameplay, see Atkins (2003).

35. Such a characterization comes close to what Salen and Zimmerman have called the "immersive fallacy," the belief that the goal of digital games, or perhaps entertainment media more generally, is to generate worlds in which the viewer or participant becomes utterly and unself-consciously absorbed (Salen and Zimmerman 2004, 450–55).

36. Poblocki and Miklaucic have both criticized the rhetoric of player omnipotence to which, they argue, *Civilization* subscribes. See Poblocki (2002,

171–74) and Miklaucic (2003, 328–34), and also see Galloway's argument that *Civilization* operates as an "allegory" of today's information society, fetishizing control while resisting traditional ideological critique (Galloway 2004).

37. McLuhan and Fiore (1967, 142).

38. In fact, Myers rejects what he calls "text-based" *and* "tech-based" positions, attributing play to a "natural-historical origin" and associating recursive replay with "basic human neurophysiology and universal cognitive practice" (2005). It is on these grounds that he draws his conclusions regarding the relative unimportance of culturally specific backstories (see below).

39. For an elaboration of this point, see Meyers (2003a), "The Attack of the Backstories (and Why They Won't Win)."

40. *Colonization* (1994), a successor of sorts to *Civilization*, in which players strive to conquer the New World, unashamedly expanded on this imperial imperative; see the closing paragraphs of Friedman (1999).

41. See also Diane Carr's response both to textual critiques of the sort leveled by Poblocki (2002) and Douglas (2002), and to Myers's exclusive focus on gameplay (Carr 2007).

42. Sid Meier has denied that the environmental aspects of the game, such as pollution and global warming, constitute any kind of political statement. A conscious effort was made, he has said, to keep the game free from any political philosophy, and that these elements were included purely for their contribution to balancing the gameplay; see Chick, Meier, and Shelley (2001). *Alpha Centauri* (1999), a narrative sequel to *Civilization*, takes up the game's story after Earth has been destroyed by war, famine, and disease, and it integrates environmental concerns more firmly still within the gameplay; see Henthorne (2003).

43. Myers argues, moreover, that within the self-contained semiotic system that is the game, signifiers assume entirely separate values from those that might ordinarily obtain in conventional social contexts. The meaning of "pollution," in other words, is wholly determined by the role that this element plays within the game (2003a, 181–82, note 5, 2005).

44. The *Civilization II Instruction Manual* warns en passant that "For game purposes, *Civilization II* treats these threats identically to industrial pollution, though in real life their effects might be considerably longer term" (Reynolds 1996, 81).

Works Cited

Aarseth, Espen J. 1997. *Cybertext: Perspectives on Ergodic Literature.* Baltimore, MD: Johns Hopkins University Press.

Abram, David. 1997. *The Spell of the Sensuous: Perception and Language in a More-Than-Human World.* New York: Vintage.

Aristotle. 1926. *The Art of Rhetoric. Aristotle in 23 Volumes.* Vol. 22. Trans. John Henry Freese. Cambridge, MA: Harvard University Press.

Atkins, Barry. 2003. "The Aesthetics of Iteration: The Plurality of Spectacle in Narrative Video Games." Presented at *DiGRA 2003 Conference: Level Up*. Utrecht, Netherlands: University of Utrecht, Digital Games Research Association. http://www.uk.geocities.com/barry.atkins3@btopenworld.com/digra.htm, accessed June 24, 2008.

Bitz, Bako. 2002. "The Culture of Civilization III." http://www.web.archive.org/web/20040324004449/http://www.joystick101.org/story/2002/1/12/222013/422, accessed June 24, 2008.

Carr, Diane. 2007. "The Trouble with Civilization." In *Videogame, Player, Text*, ed. Barry Atkins and Tanya Krzywinska, 222–36. Manchester: Manchester University Press.

Chick, Tom, Sid Meier, and Bruce Shelley. 2001. "The Fathers of Civilization: An Interview with Sid Meier and Bruce Shelley." *CG Online*. http://www.crewscut.com/index.php?title=The_Fathers_of_Civilization, accessed June 24, 2008.

Copier, Markinka. 2005. "Connecting Worlds: Fantasy Role-Playing Games, Ritual Acts, and the Magic Circle." *Proceedings of DiGRA 2005 Conference: Changing Views: Worlds in Play*. Vancouver, British Columbia. Digital Games Research Association. http://www.digra.org/dl/db/06278.50594.pdf, accessed June 24, 2008.

Crosby, Harry H., and George R. Bond, eds. 1968. *The McLuhan Explosion: A Casebook on Marshall McLuhan and Understanding Media*. New York: American Book Company.

Douglas, Christopher. 2002. "You Have Unleashed a Horde of Barbarians!": Fighting Indians, Playing Games, Forming Disciplines. *Postmodern Culture* 13:1. http://www.web.mit.edu/21w.784/www/BD%20Supplementals/Materials/UnitFour/douglasall.html, accessed June 24, 2008.

Duffy, Dennis. 1969. *Marshall McLuhan*. Toronto: McClelland and Stewart.

Friedman, Ted. 1999. "Civilization and Its Discontents: Simulation, Subjectivity, and Space." In *On a Silver Platter: CD-ROMs and the Promises of a New Technology*, ed. Greg M. Smith, 132–50. New York: New York University Press. http://www.duke.edu/~tlove/civ.htm, accessed June 24, 2008.

Galloway, Alexander R. 2004. "Playing the Code: Allegories of Control in Civilization." *Radical Philosophy* 128 (November–December):33–40. Rpt. as "Allegories of Control." In *Gaming: Essays on Algorithmic Culture*. 85–106. Minneapolis: University of Minnesota Press.

Genosko, Gary. 1999. *McLuhan and Baudrillard: The Masters of Implosion*. London: Routledge.

Gronbeck, Bruce E. 1981. "McLuhan as Rhetorical Theorist." *Journal of Communication* 31:117–28.

Havelock, Eric A. 1963. *Preface to Plato*. Cambridge, MA: Harvard University Press.

Henthorne, Tom. 2003. "Cyber-Utopias: The Politics and Ideology of Computer Games." *Studies in Popular Culture* 25:3 (April). http://www.pcasacas.org/SPC/spcissues/25.3/Henthorne.htm, accessed June 24, 2008.

Horrocks, Christopher. 2000. *Marshall McLuhan and Virtuality*. Cambridge: Icon.

Huizinga, Johan. 1955. *Homo Ludens: A Study of the Play Element in Culture.* Boston, MA: Beacon Press.

Kepes, Gyorgy. 1944. *The Language of Vision.* Chicago, IL: Paul Theobald.

Killingsworth, M. Jimmie, and Jacqueline S. Palmer, *Ecospeak: Rhetoric and Environmental Politics in America.* Carbondale: Southern Illinois University Press.

Lammes, Sybille. 2003. "On the Border: Pleasure of Exploration and Colonial Mastery in Civilization III Play the World." *Proceedings of DiGRA 2003 Conference: Level Up,* ed. Marinka Copier and Joost Raessens, 120–29. Utrecht: University of Utrecht/Digital Games Research Association. http://www.digra.org/dl/db/05163.06568, accessed June 24, 2008.

Levinson, Paul. 1999. *Digital McLuhan: A Guide to the Information Millennium.* London: Routledge.

Lister, Martin, et al., eds. 2003. *New Media: A Critical Introduction.* London: Routledge.

McLuhan, Marshall. 1958–1959. "The Electronic Revolution in North America." *International Literary Annual* 1, ed. J. Wain, 165–69. Rpt in Marshall McLuhan, *McLuhan Unbound.* Corte Madera, CA: Gingko Press.

———. 1960. "Acoustic Space." In *Explorations in Communication: An Anthology,* ed. Edmund Carpenter and Marshall McLuhan, Boston, MA: Beacon. Rpt. in *Media Research: Technology, Art, Communication: Essays by Marshall McLuhan,* ed. Michel A. Moos, 39–44. Amsterdam: G+B Arts International, 1997.

———. 1962a. "The Electronic Age—The Age of Implosion." In *Mass Media in Canada,* ed. John A. Irving, 177–205. Toronto: Ryserson Press. Rpt. in *Media Research: Technology, Art, Communication: Essays by Marshall McLuhan,* ed. Michel A. Moos, 16–38. Amsterdam: G+B Arts International, 1997.

———. 1962b. *The Gutenberg Galaxy: The Making of Typographic Man.* Toronto: University of Toronto Press.

———. 1964. *Understanding Media: The Extensions of Man.* 3rd printing. New York: McGraw-Hill.

———. 1966. "The Relation of Environment to Anti-Environment." *University of Windsor Review* 11:1 (Fall): 1–10. Rpt. in *The Human Dialogue,* ed. Floyd W. Matson and Ashley Montagu, 1–10. New York: Macmillan, 1967. Rpt. in *Media Research: Technology, Art, Communication: Essays by Marshall McLuhan,* ed. Michel A. Moos, 110–20. Amsterdam: G+B Arts International, 1997. Rpt. in Marshall McLuhan, *McLuhan Unbound.* Corte Madera, CA: Gingko Press, 2004.

———. 1969a. *Counterblast.* Designed by Harley Parker. New York: Harcourt, Brace & World.

———. 1969b. "*Playboy* Interview: Marshall McLuhan—A Candid Conversation with the High Priest of Popcult and Metaphysician of Media." *Playboy* (March): Rpt. in *Essential McLuhan,* ed. Eric McLuhan and Frank Zingrone, 233–69. New York: BasicBooks, 1995.

———. 2005. *The Classical Trivium: The Place of Thomas Nash in the Learning of His Time.* Ed. W. Terrence Gordon. Corte Madera, CA: Gingko.

McLuhan, Marshall, Kathryn Hutchon, and Eric McLuhan, 1977. *City as Classroom: Understanding Language and Media.* Toronto: Book Society of Canada.

McLuhan, Marshall, and Quentin Fiore, with Jerome Agel. 1967. *The Medium Is the Massage: An Inventory of Effects.* Harmondsworth: Penguin.

———. 1968. *War and Peace in the Global Village.* New York: McGraw-Hill.

Miklaucic, Shawn. 2003. "God Games and Governmentality: *Civilization II* and Hypermediated Knowledge." In *Foucault, Cultural Studies, and Governmentality*, ed. Jack Z. Bratich, Jeremy Packer, and Cameron McCarthy, 317–35. Albany: State University of New York Press.

Miles, David. 1996. "The CD-ROM Novel Myst and McLuhan's Fourth Law of Media: *Myst* and Its 'Retrievals.' " *Journal of Communication* 46:2 (Spring): 4–17. Rpt. in *Computer Media and Communication: A Reader*, ed. Paul A. Mayer, 307–19. Oxford: Oxford University Press, 1999.

Myers, David. 2003a. *The Nature of Computer Games: Play As Semiosis.* New York: Peter Lang.

Myers, David. 2003b. "The Attack of the Backstories (and Why They Won't Win)." In *Level Up: Digital Games Research Conference Proceedings* (CD-ROM), eds. Marinka Copier and Joost Raessens. Utrecht, the Netherlands: Faculty of the Arts, University of Utrecht. http://www.loyno.edu/%7Edmyers/F99%20classes/AttackOfTheBackstories.pdf, accessed June 24, 2008.

———. 2005. "Bombs, Barbarians, and Backstories: Meaning-Making within *Sid Meier's Civilization*." ["Bombe, barbari e antefatti. Progettazione e semantica in Civilization di Sid Meier"] In *Civilization: Storie Virtuali, Fantasie Reali*, ed. Matteo Bittanti, trans. Valentina Paggiarin, 165–83. Milan: Costa & Nolan. http://www.loyno.edu/~dmyers/F99%20classes/Myers_BombsBarbarians_DRAFT.rtf, accessed June 24, 2008.

Neve, Mario. 2004. "Does the Space Make Differences? Some Geographical Remarks about Spatial Information between Harold Innis and Marshall McLuhan." In *At the Speed of Light There Is Only Illumination: A Reappraisal of Marshall McLuhan*, ed. John Moss and Linda M. Morra. Ottawa: University of Ottawa Press.

Ong, Walter J. 1982. *Orality and Literacy: The Technologizing of the Word.* London: Methuen.

Poblocki, Kacper. 2002. "Becoming-State: The Bio-Cultural Imperialism of Sid Meier's *Civilization*," *Focaal: European Journal of Anthropology* 39:163–77. http://www.focaal.box.nl/previous/Forum%20focaal39.pdf, accessed June 24, 2008.

Reynolds, Brian. 1996. *Sid Meier's Civilization II Instruction Manual.* Hunt Valley, MD: Microprose. http://www.replacementdocs.com/download.php?view.365, accessed June 24, 2008.

Rosenthal, Raymond, ed. 1968. *McLuhan: Pro and Con.* New York: Penguin.

Salen, Katie, and Eric Zimmerman. 2004. *Rules of Play: Game Design Fundamentals.* Cambridge, MA: MIT Press.

Stearn, Gerald Emanuel, ed. 1967. *McLuhan Hot and Cool: A Primer for the Understanding of and a Critical Symposium with Responses by McLuhan.*

New York: Dial.

Stephenson, William. 1999. "The Microserfs Are Revolting: Sid Meier's Civilization II." *Bad Subjects* 45 (October) http://bad/eserver/org/issues/1999/45/stephenson.htm/, accessed June 24, 2008.

Strate, Lance, and Edward Wachtel, eds. 2006. *The Legacy of McLuhan*. Cresskill, NJ: Hampton.

Whorf, Benjamin. 1956. *Language, Thought and Reality: Selected Writings of Benjamin Lee Whorf*. Ed. John B. Carroll. Cambridge, MA: MIT Press/John Wiley.

Williams, Raymond. 1974. *Television: Technology and Cultural Form*. London: Fontana/Collins.

14

Seeing the Climate?

The Problematic Status of Visual Evidence in Climate Change Campaigning

Julie Doyle

New visual evidence of the impacts of climate change was released by Greenpeace in Patagonia today. Dramatic new photos of Patagonian glaciers taken by the research team on board the Greenpeace vessel, Arctic Sunrise, show the extent to which climate change has caused the ice to melt this century, when compared to photos of the same glaciers taken in 1928.

—Greenpeace International, "Pictures of Climate Change"

The important thing is that the photograph possesses an evidential force, and that its testimony bears not on the object but on time.

—Barthes, *Camera Lucida*

As forms of visual evidence, photographs of melting glaciers function as powerful and persuasive signs of the visible impacts of climate change upon the landscape. Indeed, such photographs have figured as vital tools in historical efforts by environmental campaign groups to bring public and political attention to the reality of climate change over the past two decades.[1] Since Greenpeace first photographed the crack in the Larsen B ice shelf in Antarctica in 1997, images of melting and retreating glaciers have been used both in campaign group literature as well as reproduced in the popular media as proof of the reality of global warming, and the resultant climate change.[2] It would appear, then, that photographs used as a means of documentary evidence function as powerful signs in the struggle against climate change, signifying the indexical proof of a world visibly scarred by a warming planet. Yet the persuasive force of these images, reliant as they are upon the referential status of the photograph as truth and reality, also illustrates the limitations of photographic representation as effective tools of communication in environmental politics. On the one hand, the visible evidence of climate change is recorded by the camera and given the status of truth by what Barthes calls the "*noeme* of photography," the referential proof that "the thing has been there" (Barthes 2000, 76). At the same time, however, the temporality inscribed in the photographic evidence of "what has been," or what Barthes identifies as the coexistence of "reality and of the past" (ibid.), proves catastrophic in the context of climate change campaigning, which necessitated action to prevent climate change *before its effects could be seen.* In other words, photographs of retreating glaciers depict an already affected environment, illustrating the current reality of climate change through the image, and at the same time signifying the failure of preventative action required to halt its acceleration. The temporal nature of climate change as an environmental issue of the greatest magnitude is thus made evident by the limited temporality of the photographic medium as a privileged form of representation within environmental campaigning.

My intention in this chapter is to examine the limitations of (documentary) photography as a discourse of evidence and truth within the context of the history of climate change campaigning. I argue that the problem of global warming, the scientific evidence of which was growing in the late eighties to early nineties, prompted a crisis of representation, and thus communication, for environmental groups. This crisis can be read in part as a consequence of the privileging of the visual within much environmental discourse; a privileging that is called into question when examined within the context of a historically "unseen" issue such as climate change. Reliance upon prediction and forecasting in the early days of climate change communication foregrounded the need for preventative action before climate change impacts could be seen. Thus it is the temporal nature of climate change that signals a crisis of the visual in environmental discourse. Here the communication of the temporality of climate change is limited by the temporality of photography as a representational medium whose evidential force is reliant upon its "certificate of presence" (Barthes, 2000, 87), which depicts "what is" or "what has been" rather than "what may be." This chapter traces the problematic relations of the visual and the temporal within environmental discourse and its visual communication, a relationship brought to the fore by the current force of comparative photographs that document the impacts of climate change upon the landscape, the consequences of which demonstrate failures to prevent climate change made evident by its currently seen status.

First, I examine the problematic history of the science of climate change and establish this as a contributing factor to the historical difficulties experienced in communicating this issue beyond the scientific community, and the role of environmental groups within this communication process. I then move on to explore the privileged role of the visual, in particular, photography, as a means of communication within environmental discourse and how this can be understood to be in conflict with the temporal and nonvisual aspect of climate change. I next discuss a specific analysis of the role of (documentary) photography within the communicative strategies of Greenpeace, more specifically, its climate change campaigning.[3] Through an analysis of current comparative photographs of glaciers produced by Greenpeace, I explore the temporal limitations of photography identified by Barthes, arguing that these limitations are brought to the fore in the specific context of climate change communication.

Establishing Climate Change: Temporal and Visual Limitations

While it would be fair to say that there is now broad scientific consensus over the reality of human-induced climate change, historically this issue has been difficult to prove as posing a real threat. One aspect of this difficulty was the "unseen" nature of climate change, in a Western culture invested in the notion that seeing is believing. A further aspect of doubt was that the science of global warming was based upon predictive modeling and forecasting rather than observable impacts (Wilson 1992). Given the investment of scientific epistemologies in the discourse of empiricism, which privileges observable "fact" over prediction and the unseen, both the science of global warming and its effective communication were limited by the discursive frameworks of scientific knowledge.[4]

From the outset, then, prediction constituted a key basis in establishing the science of global warming as the outcome of anthropogenic activity, that is, the increase in CO_2 from the burning of fossil fuels. In 1979, the WMO (World Meteorological Organization) organized the first World Climate Conference to address concerns about the effects of human activity upon the climate and to call upon the world's nations "to *foresee* and to *prevent potential* man-made changes in climate that *might* be adverse to the well-being of humanity" (IPCC, 2004, 2, emphasis added). This was followed in 1988 by the establishment of the

IPCC (Intergovernmental Panel on Climate Change) by UNEP (United Nations Environment Program) and the WMO "to provide independent scientific advice on the complex and important issue of climate change" (IPCC, 2004, Foreword). Conducting no new research, the role of the IPCC was, and is, to assess existing scientific reports published by the international scientific community in order to provide a comprehensive overview of the scientific basis of human-induced climate change. Comprised of scientists from a broad range of countries and a number of scientific disciplines, the IPCC has constituted the most authoritative scientific voice on the causes, impacts, and effects of climate change (Houghton 2004).

When the first assessment report of the IPCC was published in 1990, "an unequivocal statement that anthropogenic climate change had been detected could not . . . be made at the time" (Houghton 2004, 104). With the science still based upon prediction, in 1995, the Second Assessment Report stated, "There are many uncertainties and many factors [which] currently limit our ability to project and detect future climate change. Future unexpected, large and rapid climate system changes (as have occurred in the past) are, by their nature, difficult to predict. This implies that future climate changes may also involve 'surprises' " (IPCC, 1995, 6). However, during the mid-to-late nineties, there was increasing evidence of climate change and its impacts to support the prediction graphs and computer model simulations.

By the time of the publication of the Third Assessment Report in 2001, the causes and impacts of climate change were identified as "robust findings," while "key uncertainties" were related to "model projections" of future emissions of greenhouse gases and changes to global climate (IPCC, 2001, 30). Robust findings were related to "the existence of a climate response to human activities and the sign of the response," while key uncertainties were "concerned with the quantification of the magnitude and/or timing of the response" (ibid.). In other words, the observable signs of the existence of climate change were undeniable, while the exact magnitude of change, as well its effects, was less quantifiable. The reality and impact of climate change were thus established. The robust findings stated that the "earth's surface is warming"; that globally the "1990s is very likely [to be] the warmest decade in instrumental record"; that "atmospheric concentrations" of human-induced greenhouse gases have "increased substantially since the year 1750"; and that "most of the observed warming over [the] last 50 years [is] likely due to increases in greenhouse gas concentrations due to human activities" (IPCC, 2001, 31). By 2001, the science and reality of human-induced climate change were undeniable, with visible signs of its impacts.

It is important to acknowledge that the complexity and historical uncertainties over the science of climate change, alongside the predictive nature of its magnitude/effects, had a limiting effect on the successful communication of climate change beyond the scientific community. However, this is certainly not the only reason climate change has taken so long to establish itself in the consciousness of the public and international governments. The most obvious reason is the necessity for major economic, political, and social changes to occur on an international level so a reduction in CO_2 emissions could be achieved by moving from fossil fuel energy sources to renewables such as wind, wave, and solar. These changes impact on the everyday lifestyles and consumer habits of all citizens. However, one important means of communicating the risk and types of changes needed is through the work of NGO environmental pressure groups. Environmental groups play a key role in communicating science to the public, and thus in defining the risks and the nature of these. Analyzing the relationship between scientists, journalists, and environmental pressure groups in the mediation of climate change in three national contexts (Germany, Belgium, and France), Mormont and Dasnoy characterized the role of pressure groups "as mediators between public opinion and scientific expertise" (Mormont and Dasnoy, 1995, 56). In the context of climate change, historically an issue that was "not directly observable" and most likely accessible "via an immense scientific, technical and institutional network," environmental groups play an important role in informing the public and making the issue relevant (49). However, Mormont and Dasnoy identify the problems inherent to the effective communication of climate change as a result of the predominance of scientific fact and hypothesis upon which the issue is based.

> When it comes to climate change, the facts completely escape common experience, for it is only by communication that the issue is given meaning (as opposed to a daily occurrence such as a road accident, which ordinary experience has several ways of interpreting). The taking account of these facts (or hypotheses) by public opinion presupposes an interpretative context, one that may designate the risks and victims, in short, the social context which gives these facts meaning. (1995, 61)

Establishing interpretative frameworks could be argued to be one of the key functions of environmental campaign groups—informing and making a particular issue relevant.

How, then, to make climate change relevant, given the nature of the issue being scientifically complex, based upon predictive forecasting and model simulations? To pose this question is to foreground the more general problems of scientific empiricism, whose validity of observation/ prediction is itself limited by the methodological and epistemological frameworks through which knowledge is presented and authorized. More specific to my argument, however, the question foregrounds the limitations of environmental discourse, understood as the means by which the environment is made meaningful. In short, this necessitates examining the cultural frameworks through which we interpret the environment and environmental risks; the ways in which climate change is made meaningful as a real risk and as being relevant to the public. The lack of visual evidence and the temporal nature of climate change, two key strategies in the discursive construction of the environment, constitute key limitations for its effective communication. It is the role of environmental groups in the visual construction of the environment that I now examine.

Visualizing the Environment—Seeing Climate Change?

Above all, environment stories really need good pictures . . . global warming is very difficult because you can't actually see global warming.

—Former BBC news environment correspondent,
in Anderson, *Media, Culture, and the Environment*

Pictorially it's a tricky one to show global warming, because obviously they're showing something of the future.

—Former ITN news environment correspondent,
in Anderson, *Media, Culture, and the Environment*

Anders Hansen has argued that "environmental issues do not ordinarily articulate themselves" (Hansen, 1991, 449). The means by which environmental issues gain public and political attention thus has more to do with how the issue is articulated, or made meaningful, than the actual magnitude of the threat posed. Indeed, while climate change poses the greatest global environmental threat of modern times, this has not been matched by the extent of political will, public consciousness, or media coverage. An important reason for this lack of urgency can be examined through the nature of climate change communication, or articulation. In terms of the news media, the aforementioned epigraphs highlight how the "here" and "now" format of news discourse authorized by the visual image is in conflict with the nonvisual and temporal characteristics of global warming as an environmental issue, making it difficult to communicate and to make relevant within news format constraints (Allan 1998, 105). Much work has already been carried out analyzing media representations of the environment, calling attention to its privileging of the visual immediacy of events-based issues rather than longer-term environmental problems (Hansen, 1991, 1993; Anderson 1997; Allan 1998). While my intention here is not to focus on media representations, I do, however, want to acknowledge how the nonarticulation of climate change within the media contributes to a general interpretative framework that makes climate change meaningful (and ultimately unarticulated) through visual and temporal framing, validated by the immediacy of the image. This privileging of the visual and temporal, while characteristic of contemporary news discourse, I would argue has a longer history in relationship to environmental discourses that have themselves come to be shaped by the campaign strategies of environmental groups, whose role is to call attention to, and to define, threats to the environment. Specifically, environmental groups have come to privilege visual representations of the landscape and environment as indicative of the need for its protection. The power of the image to persuade through documentation, and to evoke an emotional response, forms a key strategy of environmental communication.

The visualization of the environment as a structuring principle of environmental discourse—how we understand, interpret, and respond to the environment—has been discussed by a number of theorists, particularly in relationship to a growing tourist industry during the nineteenth century (Urry 1990; Wilson 1992; Macnaghten and Urry 1998). Since the Enlightenment, seeing our environment has become metonymic for understanding and valuing it: a visual aesthetics and an epistemology promoted and inscribed through nineteenth-century landscape painting, photography, and, since the 1960s, satellite images from space (Ingold 1993). Visualizing the environment in order to comprehend it is a constitutive aspect of making the environment meaningful. Visual representations of the environment as a means of defining it can be taken as a "cultural given." As Hansen explains, a "cultural given" is a set of beliefs or practices of a society, through which environmental issues are made meaningful (1991, 452). One such given would be the belief in science to validate claims to authority about (environmental) risk; a distinct problem within the history of climate change science and, therefore, its effective communication. Other cultural givens are dominant ideologies, such as those about technological progress (e.g., the belief that [certain kinds of] technology will rescue the world from climate change, undermining the viability of existing technologies in the form of renewable energies).[5]

The cultural given of "seeing is believing," I would argue, is part of the means by which the environment is made meaningful, apparent in the visualizing activities of landscape painting and more recently photographic images of the beauty of our environment, used by tourism as well as environmental campaign groups. Where the cultural given of Western culture may be a belief in scientific discourse, this is intimately and historically entwined with our investment in visualizing knowledge and experiencing the world, and thus the environment, through the visual. It is not difficult to see how this investment in the visualization of the environment as a key feature of environmental discourse poses problems for the effective communication of an environmental issue such as climate change, which historically could not be seen. Indeed, besides cultural givens such as belief in science and the visual, Hansen identifies the necessity of "cultural resonances," understood as "powerful, histori-cally established, symbolic imagery," in the articulation of environmental issues, through which they gain legitimacy (Hansen, 1991, 453). Given the privileging of the landscape as a powerful symbolic image of nature since the Enlightenment, global warming and climate change have limited symbolic resonance, signifiable only when their impact has been seen on the landscape, thus effectively too late.

Photography as a specific medium of representation whose discourses articulate the notion of visual truth as well as a visual aesthetics has played a crucial role in defining the environment as visually knowable. The photographic image and photography thus occupy a privileged position within the communicative strategies of environmental campaign groups, most famously Greenpeace. Photographs are used to document environmental destruction in order to persuade the public and govern-ments to take action by their truth function, inscribed by an emotive aesthetic. Indeed, the history of photographic theory has pointed out the discursive functions of photography as truth/reality and aesthetic/creative, arguing that both coexist in all photographic images to varying degrees (Barthes, 1977a; Sekula 1982; Burgin 1982). This also can be read in terms of the signifying function, or the denotive and connotive, of all photographs (Barthes, 1977b). While more recent photographic debate often has centered on the impact of digital photography on the truth claims of analogue photography (Mitchell 1992), other theorists have pointed out that such debates deny the already contested and culturally specific nature of photographic meaning (Lister 1995; Kember 1998; Rötzer 1996). I do not intend here to enter into debates about the meaning of analogue imagery in relation to digital, and I would concur that all photographic images are contextually and culturally dependent for their meaning. More importantly, I focus on the use of the photograph as tool of documentation, which arguably persists as a dominant factor in photographic meaning, particularly in the context of environmental campaigning (Robins 1995). In this specific context, photographs still function as powerful documentary records whose purpose is political, intended to persuade governments and the public to take action to save the environment. It is for this reason that the representational limitations of photography as indexes of reality and as effective forms of political engagement need to be addressed in light of the nonvisual and temporal aspect of climate change. It is the role of photography in climate change campaigning that I will now examine.

Photographing Climate Change—The Problem of Present and Past

> Greenpeace make pictures. Pictures make Greenpeace.
>
> —Boettger, "The Role of Photography in Greenpeace's Strategy"

Greenpeace is arguably the most image-centric environmental campaign group, a view made explicit by a former picture editor who states, "Greenpeace can be regarded as an organisation with photography as its vital medium" (Boettger 2001, 12). The key feature of photography that makes it such a vital tool of communication for the organization is the notion of the photograph as a form of evidence and witness. The evidential force of the photograph is used not merely to document environmental damage but, more importantly, to capture moments of nonviolent direct action and resistance: for example, images of activists in inflatables trying to stop whaling, or activists chaining themselves to Land Rovers to call attention to the high fuel consumption of 4 × 4s. As such, the photographs document "an act of resistance, a political act" (Boettger 2001, 12). The indexical properties of the photograph, its *noeme*, are made explicit here as being fundamental to effective environmental communication.

As photographic theorists have pointed out, however, belief in the photograph as an objective record of reality is part of the discourse of photography, what Sekula calls "the established myth of photographic truth" (1982, 86). The meaning of a photograph is dependent upon an interpretation of the photographic signs which are always culturally and historically specific (Barthes 1977a, 1977b; Sontag 1977; Sekula

1982; Tagg 1982). Thus the supposed truth and objective status of the photographed landscape are reliant upon understanding the visualized landscape as representative of nature. In the context of environmental discourse, the visualized beauty of the landscape is always understood as threatened. Likewise, the activist photographed mid-action can only be interpreted as a defender of the environment and as carrying out a political act through cultural perceptions of governmental and institutional failures to protect the environment. Greenpeace assigns photography the normative role of witness and seer, subscribing to the myth of photographic truth, inscribed by the discourse of science.

If we understand one aspect of the role of environmental groups as being the communication of science and risk in an accessible manner, and photographic images as being a central means of doing this, then we can already see the problems inherent to the communication of climate change science through image. Although not reliant upon image alone to convey matters of environmental concern—indeed, Greenpeace has already been acknowledged by Mormont and Dasnoy as having played a crucial role in establishing the science of climate change through its own scientific research and publications—within the philosophy of the campaign group, photography is specifically identified as the key aspect of its public communication. As stated earlier, the crack in the Larsen B ice shelf in Antarctica was first detected and photographed by Greenpeace in 1997. Such a strong visual image of a landscape, seeming to be undeniably affected by global warming, demonstrated by the breakup in the ice sheet, proved a pivotal image in communicating the reality of global warming through its impact on the landscape. Prior to this image, Greenpeace climate change communication had focused on rising temperatures represented by images of the scorched earth as well as through computer-model simulations of a red-swathed globe to signify a warming world. Anticipated weather pattern shifts also were imaged by photographs of storm damage, hurricanes, and floods (Greenpeace International 1993, 1994; Greenpeace UK 1997). However, the photograph of the crack in the Larsen B ice shelf can be regarded as the first influential and recognizable photographic evidence of global warming. Since then the polar ice caps, alongside the world's glaciers, have become the dominant visual language of climate change impact, functioning as effective images in establishing the reality of the problem. With proclamations such as "Arctic environment melts before our eyes" and "Glacier retreats are one of the most visible and reliable signs that warming and climate change is real," information presented on the Greenpeace Web site highlights the link between seeing the impact of climate change as evidence of its reality and the central role of glacial ice as part of the symbolic landscape of climate change impacts (Greenpeace International 2002).

Problems over the effective communication of the immanency of climate change prior to 1997 can be understood in part by the lack of visible evidence of climate change, illustrated in the language of a Greenpeace publication on climate change from 1994, which stated that "the first impacts of human-induced climate change are in fact already *being felt*," but not yet seen (Greenpeace International 1994, emphasis in original). While the reality of climate change is in part identified by images of impacts, the temporality inscribed in the photograph as both present and past further problematizes the privileging of photography as an effective form of communication in the context of an environmental issue that required preventative action in the present, before future effect. Such a representational problem corresponds to the temporal conditions of the photograph identified by Barthes. Barthes's understanding of the noeme of photography as the "superimposition . . . of reality and of the past" (2000, 76) inscribed in the photographic image has a particular poignancy in the context of communicating climate change through the photograph. The temporality of climate change as both an ongoing and a future environmental condition can be said to be in conflict with the temporality of the documentary photograph as a fixed record of a particular moment in time. In other words, once climate change can be photographed, the future has been made present and authenticated as real while simultaneously being relegated to the past, a distinct problem, given the urgency with which climate change action is needed.

The current popularity of comparing historic photographic images of glaciers to those taken in the present as a means of documenting, and making real the effects of global warming on the landscape illustrates the temporality of photographic discourse and the limitations of this for environmental politics. Figure 14.1 shows one of the "[d]ramatic new photos" (Greenpeace International 2004) taken by Greenpeace of the melting Patagonian glaciers, placed underneath a photograph of the same glacier taken in 1928. Separated in time, these comparative photographs of the same glacier are intended to signify visual evidence of the effects of climate change as a result of rising temperatures. Dependent upon the indexical properties of the photograph for authentication, visible changes to the landscape (the referent) over time are made evident by the contrasting images. Thus where the photograph, unlike a moving image, captures and freezes time of a particular moment, it is this significatory power to attest to the real (moment) through which the comparative photographs gain their evidential and affective force as documentations of a changed landscape. Time, and the passing of it, is quite literally inscribed in these photographs.

Yet it is the temporal inscription and authentication of past and present through which the photographs gain their legitimacy that also

Figure 14.1. Comparative photographs of Upsala Glacier, Patagonia, Argentina. Top image taken in 1928, bottom (composite) image taken in January 2004 ©/top image: Archivo Museo Salesiano bottom image:©Greenpeace/ Daniel Beltra

represent the limitations of photographic discourse within environmental discourse and politics. These limitations become more apparent if we firstly examine the photographs in isolation. As separate images, they depict particular, and possibly different, locations. The black-and-white sepia style photograph at the top appears to illustrate a hostile environment. This panoramic photograph was taken on a geological mission, and thus it signifies within the discourse of scientific truth. Viewed on its own, the color photograph below could easily be read as an advertisement for tourism. Beautifully colored, the bright blueness of the lake and sky set against the white snowcapped mountains could depict an idyllic tourist destination. It is only when these photographs are placed together, however, that their role as documentary photographs depicting a visibly altered landscape over time gains legitimacy. Through comparison of present and past, they appear to document a process of change—change captured and made evident through retrospective viewing. Although intended as comparisons between past and present, they also can be read through the convention of before and after shots, inviting the viewer to inspect the changes that have occurred seemingly after something has happened. Given the environmental context within which these photographs gain some of their meaning, the photographs of a changed landscape due to global warming worryingly position the viewer as looking onto a landscape before climate change and after. This temporal positioning thus represents climate change as an event rather than an ongoing environmental issue, misleadingly promoting the view that climate change began after 1928 (when the top photograph was taken) rather than the reality of its development since the mid-nineteenth century. The photograph beneath this, as an image after climate change "has happened," renders redundant the potential for present and further preventative action to halt the acceleration of climate change.

Confronted by these two images of present and past, they both quite quickly appear as belonging to the past as a consequence of photographic meaning (i.e., of reality and of the past). Indeed, the top sepia-toned image, read alongside the brightly colored image below, constructs a nostalgic view of a forever-lost landscape, where the glacial ice can never be regained. As photographs intended to document environmental damage, they construct for the viewer a sense of loss, and one that appears to be irretrievable. In the context of environmental discourse promoted by Greenpeace, they function on the level of witness and seer, as well as the emotive. They appear to invite us to look at the beauty of the landscape, to experience shock, and to feel loss. The problem with this, however, is that while they may generate a feeling or an emotion through the sense of loss, they do not contribute to understanding the causes of climate change or relate them to everyday life, that is, on a level where people can actually make a difference. They say little about the future.

Another style of comparative photography used by Greenpeace in the communication of climate change impacts is shown in Figure 14.2. Here a person appears in the present landscape holding up a photograph that depicts the same landscape in the past. In this type of comparative photograph, it is more difficult to see the differences between the present landscape and that shown by the photograph held by the person. Therefore, the reality principle of this photograph is in part reliant upon authentication by the person holding the historic image. The fact that the person in the landscape, Jorge Quinteros, actually took the photograph he is holding in 1955, during an expedition sponsored by the London Royal Society, further authenticates the reality of the image, inscribed by the discourse of scientific knowledge (Greenpeace International, 2004b). At the same time, the real landscape in which Jorge stands also acts as the authentication for the truth of the photographic referent of the historic image. The indexical nature of the historic photograph and its visual documentation of an earlier time, compared to the present landscape as a means of authenticating change, is supported by the role of the person in the photograph as witness and seer, validated by his role as scientist.

Figure 14.2. Jorge Quinteros at the HPS31 glacier in Patagonia, Chile, holding the photo taken during his expedition in 1955 ©Greenpeace

In some ways, the distancing effect of the comparative photographs in Figure 14.1 is mitigated in Figure14.2 by the presence of a human within the narrative. Rather than the camera becoming the apparently unmediated eye of the landscape, as in the previous images, where the photographs become substitutes for the real landscape, here the role of the camera is made more obvious, and therefore less estranged, as a mediator between the real landscape (in which the person stands) and the photographed landscape (which the person holds). The posed style of Figure 14.2 is more familiar with personal snapshots rather than objective documentary and calls attention to the act of taking a photograph. Yet this seemingly obvious foregrounding of the subjective act of photography serves to reinscribe the very notion of photographic truth through the act of taking a photograph. As David Green and Joanna Lowry explain:

> The very act of photography, as a kind of performative gesture which points to an event in the world, as a form of designation that draws reality into the image world, is thus itself a from of indexicality. (Green and Lowry 2003, 48)

Green and Lowry argue for the performativity of photography, and thus of photographic meaning, as a positive intervention into photographic debate by identifying not only the continuing strength of "indexical inscription" but also highlighting the contextual and performative aspect of photography upon which the indexical inscription of photography is dependent. However, in the context of environmental campaigning, the photograph still functions as a powerful and persuasive form of evidence and truth. In a sense, the performative aspect of photography is found to be limited when used in the services of climate change communication.

Both sets of comparative photographs present themselves as a form of witness and documentation, inscribed by the discourse of photographic truth. They say what is, but in doing so they render climate change as a past event, captured and contained by the photographic medium. They call upon the observer to acknowledge the negative impact of climate change through visual evidence of a changed landscape, yet they do little to enable the viewer to do anything about this, given the apparent magnitude of glacial loss documented by the images. Yet such images used as testimonies of the changing landscape have become commonplace in the fight to halt the acceleration of climate change through photographing its impacts. Environmental photographer Gary Braasch began a project in 1999 committed to "The Photographic Documentation of Climate Change," within which the intention was to "repeat historic photographs to show the changes" (Braasch 2005). The repetition of images taken at specific sites in order to document and prove environmental degradation seems to have taken on an unprecedented role in current efforts to communicate the reality and rapidity of climate change, yet they remain bound by their own temporal limitations.

Conclusion: Photography as Inaction

The camera, with its insistence on perspective and the narrow field, exaggerates the eye's tendency to fragment, objectify, and estrange. Staring through a viewfinder we experience the physical world as landscape, background—the Earth as if seen from space, or as a map. At the same time, the snapshot transforms the resistant aspect of nature into something familiar, and intimate, something we can hold in our hands and memories. In this way, the camera allows us some control over the visual environments of our culture.

—Wilson, *The Culture of Nature*

Strictly speaking, one never understands anything from a photograph.

—Sontag, *On Photography*

It is difficult to deny the evidential force assigned to comparative photographs of melted glaciers, and one could argue that they have formed vital tools in the communication of climate change, making real through the visual a hitherto denied or ignored issue. The environmental and human catastrophe that threatens to be the outcome of continuing carbon dioxide emissions is represented by the photographs of already seen impacts. These photographs should alert us all to the need for action on an individual and a collective basis, yet there is something quite inactive engendered by the evidential force of these images. As Wilson (1992) explains, the very act of photography and the photographic image function to make distant an object, even if the intention is to bring it closer (see also Sontag 1977). In the context of climate change impacts, this distancing has worrisome implications, given the need for action on the everyday level. At the same time, the distancing is accompanied by an intimacy, a bringing closer, which can translate into the sense of control over that which is photographed. With the reality that climate change is not under control, the effect of these photographs may be the production of a feeling in the viewer that as long as the documentations continue, by Greenpeace or others, then that in itself is sufficient as a form of action and control.

Given the dominant ideology of Greenpeace's form of action being that of documentary witness, there is something inherently limited about this role based upon the visual (seeing translated as understanding) as well as the temporal. Adherence to the deployment of photography as an observational discourse of science is limited not only by the emphasis upon sight/seeing as truth but also the temporal limitations these pose concerning the inability to depict future scenarios as credible realities. Greenpeace's commitment to the indexical properties of documentary photography as the "here and now" marks the environment as always present or past rather than future. This is not to say that Greenpeace (or other environmental groups) does not forecast or communicate future issues. Indeed, Greenpeace was instrumental in identifying the science of climate change and forecasting future scenarios of climate change. However, the dominant visual language that forms the basis of its public communication does not enable the future to be understood in the present. The evidential force of documentary photography through which it gains its authority overrides that of simulated images that failed to alert governments and the public to what photography is now able to record and make real: visible climate change impacts. Photography cannot visualize the future as a present threat. In terms of climate change, the "what may be" is now captured as the "what is" by photography, but in doing so there is a danger that it is relegated to the past.

Acknowledgments

Thanks to the staff of Greenpeace UK, Canonbury Villas, London, for allowing me access to campaign material. Particular thanks to Angela Glienicke, Greenpeace UK picture editor, and Stokely Webster, Greenpeace UK campaigner, for the time they spent with me and the access they allowed me to images and publications. Thanks also to Emma Gibson, Greenpeace UK GM campaigner, for organizing the interviews. All views expressed in the article are my own, not Greenpeace's, unless specified. Thanks also to Irmi Karl for ongoing discussions and insightful comments upon various drafts of this chapter.

Notes

1. The First World Climate Conference took place in 1979, at which it was "agreed that human activities had increased levels of CO_2 and that more CO_2 may contribute to global warming which could have damaging consequences" (Carter 2001, 233). However, I am taking 1990 as an approximate marker here, when the first Intergovernmental Panel on Climate Change (IPCC) assessment report was produced. In the same year, Greenpeace produced a reader's guide to the IPCC reports in an effort to translate and communicate the science of global warming to a wider audience (see Greenpeace UK 1990).

2. Angela Glienicke, current Greenpeace UK picture editor, commented that the picture of a stranded walrus atop a floating piece of ice became one of the key reproduced images in the media when it was first taken in 1997 and is still requested by news media today.

3. Elsewhere, I have examined in more detail the history of Greenpeace's climate change campaigning (from 1990 to 2007) in relation to the role of the visual within its communication strategies (Doyle 2007).

4. There exists a considerable wealth of literature that examines the discursive nature of science and scientific epistemologies, ranging from feminist critiques to those of cultural historians and social scientists. It would be impossible to refer to them all here. For a selection, see Benston 1982; Foucault 2003; Haraway 1989; Harding 1991.

5. U.S. President George Bush is particularly invested in the ideology of technological progress, for example, the use of carbon capture to reduce CO_2 emissions into the atmosphere. For a discussion of the limitations of carbon capture, see Greenpeace UK 1990.

Works Cited

Allan, Stuart. 1998. "News from Nowhere: Televisual News Discourse and the Construction of Hegemony." In *Approaches to Media Discourse*, eds. Alan Bell and Peter Garret, 105–41. Oxford: Blackwell.

Anderson, Alison. 1997. *Media, Culture, and the Environment.* London: UCL Press.

Barthes, Roland. 1977a. "The Photographic Message." In *Image, Music, Text.* 15–31. London: Fontana Press.

———. 1977b. "The Rhetoric of the Image." In *Image, Music, Text.* 16–51. London: Fontana Press.

———. 2000. *Camera Lucida: Reflections on Photography.* London: Vintage.

BBC News. 2005. "In Pictures: How the World Is Changing." 2005. http://www.news.bbc.co.uk/1/shared/spl/hi/picture_gallery/05/sci_nat_how_the_world_is_changing/html/1.stm, accessed September 14, 2005.

Benston, Margaret. 1982. "Feminism and the Critique of Scientific Method." *Feminism in Canada: From Pressure to Politics,* ed. Angela R. Miles and Geraldine Finn, 47–66. Montreal: Black Rose Books.

Boettger, Conny. 2001. "The Role of Photography in Greenpeace's Strategy." In *Greenpeace, Changing the World: The Photographic Record,* ed. Conny Boettger and F. Hadman, 12–22. Steinfurt: Rasch & Rohring.

Braasch, Gary. 2005. "World View of Global Warming: The Photographic Documentation of Climate Change." http://www.worldviewofglobalwarming.org/pages/background.html, accessed 14 September 2005.

Burgin, Victor, ed. 1982. *Thinking Photography.* London: Macmillan Press.

Carter, Neil. 2001. *The Politics of the Environment: Ideas, Activism, Policy.* Cambridge: Cambridge University Press.

Doyle, Julie. 2007. "Picturing the Clima(c)tic: Greenpeace and the Representational Politics of Climate Change Communication." *Science as Culture* 16:2 (June):129–50.

Foucault, Michel. 2003. *The Order of Things: An Archaeology of Human Sciences.* London: Routledge.

Glienicke, Angela. 2004. Personal interview. December 9, 2004.

Green, David and Joanna Lowry. 2003. "From Presence to the Performative: Rethinking Photographic Indexicality." In *Where Is the Photograph?,* ed. David Green. 46–60. Brighton & Kent: Photoforum and Photoworks.

Greenpeace UK. 1990. *Climate Change: A Reader's Guide to the IPCC Reports.* London: Climate Action Network.

———. 1997. *Putting the Lid on Fossil Fuels: Why the Atlantic Should Be a Frontier against Oil Exploration.* London: Greenpeace UK.

———. 2005. *Climate Change and the G8 Summit.* http://www.greenpeace.org.uk/MultimediaFiles/Live/FullReport/7133.pdf, accessed September 14, 2005.

Greenpeace International. 1993. *Emerging Impacts of Climate Change? How Lucky Do You Feel?* Amsterdam: Greenpeace International.

———. 1994. *Climate Time Bomb: Signs of Climate Change from the Greenpeace Database.* Amsterdam: Greenpeace International.

———. 2002. "Arctic Environment Melts Before Our Eyes." http://www.greenpeace.org/international/news/glaciers-melt-before-our-eyes, accessed September 14, 2005.

———. 2004a. "Pictures of Climate Change from the Disappearing Glaciers of Patagonia." http://www.greenpeace.org/international/press/releases/pictures-of-climate-change-fro#, accessed March 7, 2007.

———. 2004b. "Where Has All the Ice Gone?: Patagonia Revisited." http://www.greenpeace.org/international/news/patagonia-revisited, accessed March 7, 2005.

Hansen, Anders. 1991. "The Media and the Social Construction of the Environment," *Media, Culture and Society* 13:443–56.

———. 1993. "Greenpeace and Press Coverage of Environmental Issues." In *The Mass Media and Environmental Issues*, ed. Anders Hansen, 150–78. Leicester University Press.

Haraway, Donna. 1989. *Primate Visions: Gender, Race, and Nature in the World of Modern Science*. London: Routledge.

Harding, Sandra. 1991. *Whose Science? Whose Knowledge? Thinking from Women's Lives*. Ithaca, NY: Cornell University Press.

Houghton, John. 2004. *Global Warming: The Complete Briefing*. 3rd ed. Cambridge: Cambridge University Press.

Ingold, Tim. 1993. "Globes and Spheres: The Topology of Environmentalism." In *Environmentalism: The View from Anthropology*, ed. Kay Milton, 33–42. London: Routledge.

IPCC. 1995. *Second Assessment Climate Change*. http://www.ipcc.ch/pub/sa(E).pdf, accessed August 2, 2005.

———. 2001. *Climate Change Synthesis Report: Summary for Policymakers*. http://www.ipcc.ch/pub/un/syreng/spm.pdf, accessed August 2, 2005.

———. 2004. *16 Years of Scientific Assessment in Support of the Climate Convention*. http://www.ipcc.ch/about/anniversarybrochure.pdf, accessed August 2, 2005.

Kember, Sarah. 1998. *Virtual Anxiety: Photography, New Technologies and Subjectivity*. Manchester: Manchester University Press.

Lister, Martin. 1995. "Introduction." In *The Photographic Image in Digital Culture*, ed. Mitchell Lister, 1–28. London: Routledge.

Macnaghten, Phil, and John Urry. 1998. *Contested Natures*. London: Sage.

Mitchel, W. J. 1992. *The Reconfigured Eye: Visual Truth in the Post Photographic Era*. MIT Press.

Mormont, Marc, and Christine Dasnoy. 1995. "Source Strategies and the Mediazation of Climate Change." *Media, Culture and Society* 17: 49–64.

Robins, Kevin. 1995. "Will Image Move Us Still?" *The Photographic Image in Digital Culture*, ed. Martin Lister, 29–50. London: Routledge.

Rötzer, Florian. 1996. "Re: Photography." In *Photography after Photography: Memory and Representation in the Digital Age*, ed. Hubertus v. Amelunxen, Stefan Iglhaut, and Florian Rötzer, in collaboration with Alexis Cassel and Nikolaus G. Schneider, ___. Amsterdam: G& B Arts.

Sekula, Alan. 1982. "On the Invention of Photographic Meaning." In *Thinking Photography*, ed. Victor Burgin, 452–73. London: Macmillan.

Sontag, Susan. 1977. *On Photography*. New York: Dell.

Tagg, John. 1982. "The Currency of the Photograph." In *Thinking Photography*, ed. Victor Burgin, 110–41. London: Macmillan.

Urry, John. 1990. *The Tourist Gaze: Leisure and Travel in Contemporary Societies*. London: Sage.

Webster, Stokely. 2004. Personal interview. December 9.

Wilson, Alexander. 1992. *The Culture of Nature: North American Landscape from Disney to Exxon Valdez*. Oxford: Blackwell.

Wilson, Kris. M. 2000. "Communicating Climate Change through the Media: Predictions, Politics, and Perceptions of Risk." In *Environmental Risks and the Media*, ed. Stuart Allan, Barbara Adam, and Cynthia Carter, 201–17. London: Routledge.

15

Afterword

M. Jimmie Killingsworth and Jacqueline S. Palmer

Ranging from the sublime and apocalyptic to the scientific and mundane, images have played a powerful and controversial role in global ecopolitics since the first Earth Day in 1970. Consider a few prominent examples that play at best a minor role in the foregoing chapters:

- The great Yosemite photographs of Ansel Adams, originally dating back to the 1930s and 1940s, which became a standard feature in the publications of the Sierra Club (an odd match in many ways, as James Frost has noted—Adams's modernist formalism, an aesthetic conventionally opposed to ideological signification, applied effectively to political ends);

- The tearful Native American in traditional dress surveying scenes of contemporary waste on the old television ad (later discredited in Shepard Krech III's troubling book *The Ecological Indian*);

- The repeated till readily recognized shot of the nuclear reactor at Three Mile Island, an ominous (though likely a harmless) wisp of steam rising from the tower;

- The heroic (if carefully staged) interposing of the Greenpeace commandos' vulnerable human flesh afloat on nothing but rubber and air between Russian harpoon cannons and their threatened prey, the whales of the northern seas; and

- The myriad scenes of environmental destruction that have permeated the news and entertainment media over the past three decades: seabirds coated with black grime after oil spills, clear-cut forests, medical waste strewn on beaches, eroded deserts or wasted farmlands with starving brown faces in the foreground, the cityscapes of a *Blade-Runner*-like future enacted in the present, the

ubiquitous smoke, smog, bubbly brown water foaming through
pipes, Styrofoam and plastic trash bobbing offshore, mountainous heaps of steaming garbage, animals deformed and mutilated,
and the urban and suburban waste landscapes—horizon-striving
traffic jams and deserted strip shopping centers with pavement
cracked and windows broken.

From the vast scenes of the wilderness imagination—what Sierra Club
activist David Brower called "scenic climaxes" in his conversations with
John McPhee—to the nightmares of a lifeworld rendered unfit for human
habitation, the rhetoric of environmentalism has been built upon mass-
mediated imagery. If anything, the visual is "privileged" in environmental
communication despite its limitations, as Julie Doyle suggests in her
chapter. It is too often a manifestation of the close-to-pornographic
obsession with looking, as revealed in the trenchant analysis of Bart
Welling in these pages. It appeals rhetorically to the insatiably hungry
eyes of touristic and consumerist cultures. And yet until this volume,
the visual aspects of environmental rhetoric have only rarely attracted
scholarly attention.

     There may be good reasons for this neglect. Predominantly liter-
ate and discursive, scholarly imagination, let us hope, tends to resist
the reductive power, the glibness and seeming immediacy (*im-mediacy*,
suggesting a lack of mediation) in the glossy picture. Consider a recent
statement by Republican pollster Frank Luntz in a discussion of his
book *Words That Work: It's Not What You Say, It's What People Hear*
on NPR's *Fresh Air*. According to Luntz, if a "majority of ordinary
Americans" are shown a picture of a landscape after oil companies have
drilled exploratory wells and say that "exploration" is a better word than
"drilling" for what they seem to see, then who are we to say that they
are wrong? In other words, if it looks okay, it is okay. But pictures can
never tell the whole story, certainly not still pictures that can narrate
only in series (as Roland Barthes compellingly argued decades ago; see
Killingsworth and Gilbertson 1992, chapter 3, on signs and sequencing).
And uninformed perceptions of a limited visual reality cannot become
the foundation for public deliberation, can they? From the perspective of
Ecosee, the issue here is not a matter of competing images or who has
the most compelling image but rather what the picture leaves out. Does
it tell the whole story, or tell a story at all, and if not, who gets to say
what the picture means or narrate the story it illustrates? Is the story
a past event with no tangible connection to current or future human
actions? As Sean Morey notes, the institutions that "control what the
images 'say,' the pictorial manipulators that give picture-speech acts their

illocutionary force and perlocutionary effects, are hegemonic structures that determine what gets shown" and what gets said about the images that are selected for showing.

Two words often applied to successful images are "stunning" and "arresting," both of which involve the cessation or truncation of action. Stopping can be a good thing in environmental politics. It is good to stop pollution, species depletion, and environmental injustice. But as we suggested in our 1992 book on environmental rhetoric, to stop thinking or to stop discussion can be one effect of a prominent trend in the discourse of environmental activism and its opposition. The oversimplification of public language that we called "ecospeak" (mimicking the Orwellian concept of *Newspeak* in *1984*, the replacement of thoughtful discourse with slogans and other linguistic formulas) can lead to the reduction of options into stark opposites such as clean air versus good jobs, environmental protection versus economic development, and life as we know it versus a return to living in caves. Images with their power to stun and arrest can contribute to this cessation of thought and the binding of action that comes from the perception of choices as unacceptable, of problems as intractable. One could not find a better example of ecospeak than the division inherent in the quotation from Robert Gottlieb, given in the Introduction to this book: "Environmentalism has become associated with compelling ideas and images—whether Nature (the value of wilderness) or Society (the negative associations of urban pollution or hazards)." The photographer's selectivity of composition, along with the airbrush and computer enhancement, can purify this division even more effectively than verbal imagery could ever hope to do. But then how does the experience of viewing the pairing of wilderness and urban photographs jibe with the experience of the *lifeworld*, the sum total of the phenomenological experience of the world? Rhetorical criticism often involves the restoring of context that the framing of the photographs has eliminated. Further imagery may help in such an endeavor, but resistance to the image, urged by the hermeneutics of suspicion, is the starting point.

Ultimately, however, neglect will not suffice as a form of resistance to the reductive power of imagery, for neglect also is a form of cessation and truncation. To deal with pictures, however, we may need more words, not fewer. The cliché that a picture is worth a thousand words, alluded to more than once in these pages, might suggest the power of images to replace verbal expression, but it also might suggest the opposite—that every picture demands a thousand words if it is to become meaningful, a perspective nicely realized in the meticulous and theoretically extended analysis of these chapters. Against the power of the image to reduce or replace discourse stands the verbal copiousness realized first of all in the multiplication of words, then in the making of new words—*ecosee* in the title and the outpouring of neologisms and compounds that follows (the championship going to the central term of Cary Wolfe's provocative chapter: *carnophallogocentrism*). While we might frown on proliferations of jargon in other contexts, here we cannot. New language becomes a sign of a struggle to gain new perspectives, new understandings. It opposes the reduction of language in *Newspeak* and *ecospeak*.

The antireductive impulse also is served by the expansion of rhetorical investigation beyond the verbal, the need to create a visual rhetoric that is at the heart of this book. While visuality itself can be reductive or abusive—as indicated by the endless repetitions today of the poststructuralist critique of "The Gaze," a virtual scopophobic reaction to the scopophilia of Western visual culture—a visual rhetoric must remind us that the effect of imagery itself depends upon the attitudes of producers and users of the images and is not inherent in the medium itself. As Steve Baker's chapter argues, "to look" or "to watch" is not the only possible response to visual experience; other nuances of attitude come into play with conceptualizations such as "to observe" or "to bear witness," or "to respond viscerally, with emotion or even physical revulsion" (responses that do not necessarily reduce to the erotic or lustful gaze). Following Wendy Wheeler, Baker prefers "to attend." Hinting at a form of being fully present to experience—attending to the actual presence of animals in the manner of the artists whom Baker studies, Olly and Suzi, or attending to the artists themselves in Baker's own interviews and close analyses of their works—the visual experience overlaps with the aural. "Attending" has the root meaning of "to listen" (as does *audience*, perhaps the most crucial common term in all rhetorical analysis). In this sense, face and voice become reunited. The fullest experience of the visual becomes "being there" (as in the old Peter Sellers film based on Jerzy Kosinski's novel, not to mention Heidegger's *Dasein*), the complete emplacement of the phenomenological body, or the embodiment of all experience, including mental experience as well as sight and hearing.

The goal of complete involvement takes us away from the visual experience of "ecoporn." While joining Bart H. Welling in acknowledging the continuity between pornography and other forms of photographic and cinematic performance, scholarship in visual rhetoric can seek alternative attitudes and responses. The first step is to match the kind of terminological multiplicity and theoretical diversity we find in these chapters with the equally impressive multiplication of examples and types of imagistic experience. Eduardo Kac's "polysemic" art, the media immersion of electronic gaming, the personal or activist Web site, calendars, postcards, nature films, museums, national parks—all provide alternatives to mass-media imagery and "image events" that have preoccupied environmental rhetoricians to date (see DeLuca 1999, for example).

Beyond the further search for alternatives and the deeper analysis of the images at hand, where do the doors opened by this collection lead us? A few possible directions follow:

- *A richer look at historical practices of visual rhetoric* on the model provided by such scholars as Wickliff and Clark, Halloran, and Woodford. The histories of painting, photography, and film might be particularly relevant, along with work on the emergence of the new media (following Kittler 1990 and Manovich 2001).

- *A more explicit revision of rhetorical theory based on insights from the study of the visual and ecological imagination.* Ecosee authors hint at a substitution of image for ethos, for example. We also might consider the effect of photographic vision on the rhetoric of place (a direction already pursued in some ecopoetical work—the treatment of Whitman's scenic imagination in his late-life nature writing in Killingsworth's *Walt Whitman and the Earth*, for example). We need more work on the power of the visual in argumentation (the initial pedagogy for which has been outlined in *Picturing Texts* by Faigley, George, Palchik, and Selfe). We need to know the power of the visual in affecting the rhetoric of place (see Killingsworth, *Appeals in Modern Rhetoric*, chapter 5). And we need to take W. J. T. Mitchell's work to the next step in considering the way the visual imagination invades the language of ecological speech and writing, the interplay of ecospeak and ecosee (see, for example, Killingsworth 2006 for an initial critique of the way map metaphors play out in ecocriticism).

- *Consideration of the deep differences between still imagery and film,* which may be too easily lumped together when we think of the "visual" as opposed to the "verbal," a necessary first step in devising a visual rhetoric. Some good points of departure for this work are suggested in David Blakesley's (2003) interesting collection on the rhetoric of film, *The Terministic Screen*.

- *A fuller study of the interchange among genres and discourse communities, especially among journalistic, artistic, and technical images.* The authors of *Ecosee* provide a needed foray into the fine arts. Now we need more work on the use of technical visuals in ecological discourse, such as charts, graphs (which can be used, for example, to deal with the issues of temporality raised here

by Julie Doyle), and especially maps. An outstanding example
of the mixed-media discourse pursued on television, the Web,
and the wide screen of feature filming is the surprisingly successful treatment of global warming in Al Gore's *An Inconvenient
Truth* (the movie), a kind of Power Point slide show on steroids
and a watershed for ecopolitical imagery, blending the scientific
and technical with the activist and scenic, and melding the
genres of the campaign ad, the nature film, and the journalistic
exposé.

- *More work on the rhetoric of production* to match the excellent
  start that the *Ecosee* authors make in critical rhetoric. Because
  rhetoric traditionally involves not only the critique but also the
  production of discourse, we need fuller insights into the complexities of image production.

Allow us to expand a bit on this last point. One model for getting inside the processes of production appears in Steve Baker's interviews with Olly and Suzy. Another is suggested in Quinn Gorman's
explorations into the question of how we come to see things as we
do. Following Wolfe's thinking on artistic and rhetorical encounters
with the animal other, we would like to pause over the question of
rhetorical production a bit longer in a brief meditation that focuses
on our efforts to produce a powerful but tolerable discourse on the
phenomenon of roadkill.

The topic came to fore for us in a recent project that involved
producing text to go with a book of photographs about the natural world
of our home region, the middle Brazos Valley of Texas (see Killingsworth
and Steele forthcoming). Any account of how we experience the native
fauna would be incomplete without acknowledging that people living
in suburban and rural Texas are most likely to encounter wild animals
(other than birds) as corpses on the byways of human transportation.
The idea did not arise merely from observation and listening to people
talk but was mediated by a rich intertextual and contextual experience.
Roadkill figures widely in jokes—visual as well as verbal, such as the
frequently reproduced postcard or poster showing an armadillo or a
possum painted over with a double yellow line on the highway, as if
the road workers did not bother to pause for roadkill; or Gary Larson's
cartoon of a spatula-wielding bird approaching a splattered critter on a
road over the caption "Tools of the Common Crow." There are references in popular music—such as Greg Brown's metaphor about roadkill

on the information highway in his song "Slant Six Mind"—and bands who use roadkill in their names or song titles. There are fine essays and poems about roadkill—Barry Lopez's "Apologia," William Stafford's "Traveling through the Dark," and Gary Snyder's "The Dead by the Side of the Road," to name only three of the most moving examples. There is even a field guide to identifying crushed creatures—Knutson's *Flattened Fauna.*

Out of this increasingly rich tradition came a number of verbal segments for the Brazos Valley essay, reflecting on the infringement of suburban development on habitat, the interchange of diurnal and nocturnal life, and the simple shock of coming face-to-face with dead and dismembered animals on a daily basis. The photographer for the project, anthropologist Gentry Steele, appreciated the written treatment of the topic and, like us, found roadkill alternately disturbing and fascinating. He told us that years ago he came to be known as "the roadkill king" among his graduate students, because he encouraged them to collect dead things for use in his archaeology labs and courses on animal anatomy. He would sometimes come home to plastic bags of roadkill on his doorstep. Our editors and prepublication reviewers likewise found the roadkill passages compelling if somewhat unusual for the genre in which we were working—the regional nature book, in which the text loosely supports or augments the dominant attraction of the photographs. In this case, the roadkill segments definitely augmented or supplemented the photographs. No one ever suggested that Steele add some photographs of animals sacrificed to the gods of human mobility. In the spirit of the investigations in *Ecosee*, we might wonder, why not? Why would visualization prove even more rhetorically risky than references in the text?

Beyond the revulsion we might expect from the ordinary reader of coffee-table nature writing—the yuck factor—we could predict a critically edgier and more intellectually rigorous, but no less offended, response from opponents of ecoporn such as Bart Welling should we have decided to portray the horrific dismemberment and bloody mess of roadkill in full-color, straight-on photography (which you can find on the Web, along with the rest of the porn). We would be appealing to the side of visual experience that caters to morbid curiosity and the highway experience of rubbernecking at accident sites. Of course, not all roadkill is disgusting. Occasionally you find a clean kill (loving pictures of which also are available on the Web). But could the photograph of such an object really embody the phenomenon of roadkill (as depicted in literature) as a smeary offense against nature, a wasteful by-product of mechanized civilization, the explosive confrontation of the animal body with overpowering technology, the intersecting of superhighways with the ancient pathways of deer, raccoon, possum, javelina, armadillo, coyote, rat, squirrel, tarantula, owl, and rattlesnake? Clearly a productive dilemma confronts the rhetorical reproduction of roadkill as a visual experience.

The best result we might offer is represented in Figure 15.1, a photograph made during the formative time of the Brazos Valley book but not intended specifically for use in the project.

The image represents the discarded remains of a small dog that we encountered on a river access road leading under a bridge over the Brazos River near Marlin, Texas. Here is the text (by Killingsworth) that resulted from the encounter and actually made it into the book:

> The water and the wind seem to be the main animating forces on this winter day. I see wisps of dirty gray fiber fluttering in the breeze around the desiccated corpse of a skunk and stuck in stiff blades of winter grass—signs of the lingering cotton trade in these parts left by trucks carrying product to market. On the little road down to the river, we find the almost perfectly preserved skeleton of a small dog, dumped here dead no doubt, but flattened by subsequent traffic and picked clean by

Figure 15.1. *Roadkill* (photograph by Jacqueline S. Palmer)

crows and other carrion-eaters. In an earlier stage of decay is
a goat beneath the bridge. I've found animal remains on river
access roads all over Texas. I guess the idea is to dump the dead
things and hope the river will carry them someplace else—the
afterlife maybe.

The photograph, which until now has gone unpublished, nicely
illustrates the slightly ironic magical realism of the prose with its tran-
scendental undercurrent. But it could never stand for anything like a
typical roadkill encounter—it is too clean and neat. Infrequently used,
the road on which we found the corpse allowed it to rot in peace in the
prevailingly dry air and be pecked clean by undisturbed vultures, crows,
and fire ants (the carrion feeders who frequently become roadkill victims
themselves on busier highways). The skeleton and bits of fur that remained
were never scattered to the four winds by the rush of interstate traffic.
They were never washed away or spirited off to landfills by a suburban
cleaning crew. They were left to desiccate on the macadam of a one-lane
back road, forming an image that reminded us first of a lab specimen and
later, when viewed as a photographic text, of a cave painting or some
other fetish of a hunting people who might have themselves picked the
creature clean of meat and organs before contemplating and reproducing the formal beauty of its raw remains. In short, the more we looked
at it, and with the distance of time, imagination, and representational
technology, the less it seemed like roadkill. Rather, in this eco(see)stem,
it is a poignant reflection of different species (Homo sapiens included)
interacting over time in a natural and predictable way. It tells us phenomenal truths about who we are—and who we are not.

Images are but one part of phenomenal systems that include talk,
writing, history, myth, technology, walks in the open air with beloved
companions, and encounters with animals live and dead. To understand the things that our eyes present to us, we cannot study them
outside the fullest possible recounting of *experience*—the truest object
of theory and practice in rhetoric, the environmental imagination, and
critical ecology.

Works Cited

Barthes, Roland. 1977. *Image, Music, Text*. Trans. Stephen Heath. New York:
    Hill and Wang.
Blakesley, David, ed. 2003. *The Terministic Screen: Rhetorical Perspectives on
    Film*. Carbondale: Southern Illinois University Press.

Clark, Gregory, S. Michael Halloran, and Allison Woodford. 1996. "Thomas Cole's Vision of 'Nature' and the Conquest Theme in American Culture." In *Green Culture: Environmental Rhetoric in Contemporary America*, ed. Carl G. Herndl and Stuart C. Brown, 213–35. Madison: University of Wisconsin Press.

DeLuca, Kevin Michael. 1999. *Image Politics: The New Rhetoric of Environmental Activism*. New York: Guilford.

Faigley, Lester, Diana George, Anna Palchik, and Cynthia Selfe. 2004. *Picturing Texts*. New York: Norton.

Frost, James. 2000. "Modernism and the New Picturesque." In *Technical Communication, Deliberative Rhetoric, and Environmental Discourse*, ed. Nancy W. Coppola and Bill Karis, 113–38. Stamford, CT: Ablex.

Gottlieb, Robert. 2001. *Environmentalism Unbound: Exploring New Pathways for Change*. Cambridge, MA: MIT Press.

Killingsworth, M. Jimmie. 2004. *Walt Whitman and the Earth: A Study in Ecopoetics*. Iowa City: University of Iowa Press.

———. 2005. *Appeals in Modern Rhetoric: An Ordinary Language Approach.* Carbondale: Southern Illinois University Press.

———. 2006. "Maps and Towers: Metaphors in Ecological Discourse." *ISLE: Interdisciplinary Studies in Literature and Environment* 13:1:83–89.

Killingsworth, M. Jimmie, and D. Gentry Steele. Forthcoming. *Brazos Valley Reflections*. College Station: Texas A&M University Press.

Killingsworth, M. Jimmie, and Jacqueline S. Palmer. 1992. *Ecospeak: Rhetoric and Environmental Politics in America*. Carbondale: Southern Illinois University Press.

Killingsworth, M. Jimmie, and Michael K. Gilbertson. 1992. *Signs, Genres, and Communities in Technical Communication*. Amityville, NY: Baywood.

Kittler, Friedrich A. 1990. *Discourse Networks 1800/1900*. Trans. Michael Metteer, with Chris Cullens. Stanford, CA: Stanford University Press.

———. 1999. *Gramophone, Film, Typewriter*. Trans. Geoffrey Winthrop-Young and Michael Wutz. Stanford, CA: Stanford University Press.

Knutson, Roger M. 1987. *Flattened Fauna: A Field Guide to Common Animals of Roads, Streets, and Highways*. Berkeley, CA: Ten Speed Press.

Krech, Shepard, III. 1999. *The Ecological Indian: Myth and History*. New York: Norton.

Lopez, Barry. 1999. "Apologia." In *Literature and the Environment: A Reader on Nature and Culture*, ed. Lorraine Anderson, Scott Slovic, and John P. O'Grady, 75–79. New York: Longman.

Luntz, Frank. 2007. *Words That Work: It's Not What You Say, It's What People Hear*. New York: Hyperion.

Manovich, Lev. 2001. *The Language of New Media*. Cambridge, MA: MIT Press.

McPhee, John. 1971. *Encounters with the Archdruid*. New York: Farrar.

Mitchell, W. J. T. 1994. *Picture Theory: Essays on Verbal and Visual Representation*. Chicago, IL: University of Chicago Press.

Snyder, Gary. 1974. "The Dead by the Side of the Road." In *Turtle Island*. New York: New Directions.

Stafford, William. 1999. "Traveling through the Dark." In *Literature and the Environment: A Reader on Nature and Culture*, ed. Lorraine Anderson, Scott Slovic, and John P. O'Grady, 79–80. New York: Longman.

Wickliff, Gregory A. 2000. "Geology, Photograph, and Environmental Rhetoric in the American West of 1860–1890." In *Technical Communication, Deliberative Rhetoric, and Environmental Discourse*, ed. Nancy W. Coppola and Bill Karis, 77–111. Stamford, CT: Ablex.

Contributors

Steve Baker is professor of art history at the University of Central Lancashire, United Kingdom. He is the author of *The Postmodern Animal* and of *Picturing the Beast: Animals, Identity, and Representation*, and is a member of the editorial board of *Society & Animals*. He is also a founding member of the UK Animal Studies Group (whose jointly authored book *Killing Animals* was published by Illinois University Press in 2006), and a contributor to Berg's 2007 collection *The Animals Reader*. He is currently working on a book entitled *Art before Ethics: Creativity and Animal Life*.

Pat Brereton is the associate dean of research in the faculty of humanities and social science at Dublin City University in Ireland, where he also lectures in film and media studies at both undergraduate and postgraduate level. He is particularly interested at present in new media literacy having edited a volume of *Convergence* (vol. 13, no. 2, May 2007), on DVD add-ons. His books include a *Historical Encyclopaedia of Irish Cinema* with Dr. Roderick Flynn (Scarecrow Press, 2007) and *Hollywood Utopia: Ecology in Contemporary American Cinema* (Intellect Press, 2005). This latter work emanated from his PhD study and foregrounds an exploration of ecology and nature as represented on film. Ecology as a focus of research also permeates several articles recently, including "Nature Tourism and Irish Film" in *Irish Studies Review* (2006).

Heather Dawkins is an art historian and the author of *The Nude in French Art and Culture, 1870–1910*, published in 2002. Her research has been published in the journals *Art History, Differences: A Journal of Feminist Cultural Studies,* and *RACAR*, as well as the anthology, *Dealing with Degas: Representations of Women and the Politics of Vision.* Dr. Dawkins is an associate professor and an associate dean in the faculty of arts and social sciences at Simon Fraser University in British Columbia, Canada.

Teresa E. P. Delfín is assistant professor of anthropology at Rhodes College in Memphis, Tennessee, where her areas of expertise are Andean anthropology, gender and sexuality, anthropology of the environment, indigenous social movements, and rhetoric. She has a PhD in modern thought and literature from Stanford University.

Sidney I. Dobrin is an associate professor in the Department of English at the University of Florida. He is author and editor of more than thirteen books.

Julie Doyle is a senior lecturer in media and communication studies at the University of Brighton, United Kingdom. Her research interests include environmental politics and climate change communication and the cultural history of surgery in constructions of gender, sexual identity, and the body. She has published work on the visual politics of climate change communication and on surgery and embodiment, in journals such as *Science as Culture, Social Semiotics,* and *Women: A Cultural Review.*

Kathryn Ferguson is an Australian Research Council Postdoctoral Fellow at the Australian Research Council Center of Excellence. She is a cultural historian who investigates the means and mechanisms through which spaces and places accrue various cultural meanings and different social valuations over time. Her current research is an investigation of the public history of the Great Barrier Reef.

Quinn R. Gorman received his MA in literature and environment at the University of Nevada, Reno, and his research examines the intersection of rhetorical theory and environmental politics. He has taught classes at the University of Georgia and, as a visiting lecturer, at Babes-Bolyai University in Cluj-Napoca, Romania. He currently resides in Baltimore, Maryland, and works in the public sector.

M. Jimmie Killingsworth is professor and head of the Department of English at Texas A&M University. He is the author of many books and articles on environmental rhetoric and ecocriticism, including *Walt Whitman and the Earth: A Study in Ecopoetics* (2004) and, with Jacqueline Palmer, *Ecospeak: Rhetoric and Environmental Politics in America* (1992).

Sean Morey teaches writing and digital media in the Department of English at the University of Florida. His research focuses on Gregory L. Ulmer's theories of *electracy* and addresses the intersection between networks, systems theory, and composition as well as the role of visual rhetoric in constructions of environment. He also designs and maintains the Ichthyology Division's Web site at the Florida Museum of Natural History, including the Web site for the International Shark Attack File (ISAF) and various other shark conservation organizations.

Eleanor Morgan is an artist currently living and working in London, England. She has exhibited in Europe and North America. Her work explores the relationship between nature and culture, and she attempts to create art that hovers between the two. This has included sculpting with spider's silk, encouraging ants to draw self-portraits, and filming a close encounter with a giant green sea anemone.

Simone Osthoff is Associate Professor of Critical Studies at the Pennsylvania State University. She holds a PhD in Media and Communications from the European Graduate School in Switzerland, an MA degree in Art History, Theory, and Criticism from the School of Art Institute of Chicago, and an MFA degree from the University of Maryland. She is a member of the *Leonardo Review* panel since 2000. Her many interviews, essays, and book chapters focusing upon contemporary art have been translated into eight languages and included in among others, various MIT Press and Routledge books as well as in international art magazines and web publications. She received a Fulbright Fellowship in 2003, and is a frequent lecturer in the United States and abroad having participated in dozens of conferences worldwide.

Jacqueline S. Palmer, EdD, teaches technical writing, editing, and Web writing at Texas A&M University in College Station. She has publications on assessment, environmental rhetoric, and technical communication, including *Ecospeak: Rhetoric and Environmental Politics in America*, coauthored with M. Jimmie Killingsworth.

Spencer Schaffner received his doctorate from the University of Washington in composition rhetoric and is currently assistant professor of English in the Center for Writing Studies at the University of Illinois, Urbana-Champaign. Schaffner is working on a book manuscript about North American field guides to birds, a project that explores how bird-watchers modify their guides to make them personally and functionally more desirable.

Tom Tyler is a senior lecturer in communication, media, and culture at Oxford Brookes University, United Kingdom. His current research interests include media ecological approaches to communication technologies, especially digital games, and the uses to which inhuman animals have been put by philosophy and cultural theory.

Bart H. Welling is an assistant professor of English and an Environmental Center Fellow at the University of North Florida in Jacksonville, where he teaches a range of courses in modern U.S. literature and film, environmental literature and culture, ecocriticism, and animal studies. He is presently working on a number of research projects dealing with linguistic and visual representations of human-nonhuman relationships, along with an interdisciplinary Environmental Conflict Resolution Exercise offered through UNF's Environmental Center.

Cary Wolfe is the Bruce and Elizabeth Dunlevie Professor of English at Rice University. He has written widely on contemporary theory and U.S. culture in *diacritics, boundary 2, Cultural Critique, New German Critique,* and many others. His books include *Critical Environments: Postmodern Theory and the Pragmatics of the "Outside"* (1998), the edited collection *Zoontologies: The Question of the Animal* (2003), and *Animal Rites: American Culture, the Discourse of Species, and Posthumanist Theory* (2003).

Index